To My Father
Ian Dowie

The world has left the Earth behind it.
John Berger, Pig Earth *(1979)*[1]

Contents

Preface Courting Irrelevance ix

Acknowledgments xv

Introduction The Polite Revolution 1

1 The Environmental Imagination 9

2 The Culture of Reform 29

3 Fix Becomes Folly 63

4 Antagonists 83

5 The Third Wave 105

6 Environmental Justice 125

7 Faded Green 175

8 The Fourth Wave 205

Epilogue A Reason (or Two) for Hope 259

Notes 265

Appendix A Principles of Environmental Justice 284

Appendix B Sierra Club Centennial Address,
Michael L. Fischer 286

Appendix C The Wise Use Agenda 293

Index 299

Preface

Courting Irrelevance

When the history of the twentieth century is finally written, the single most important social movement of the period will be judged to be environmentalism.
Robert Nisbet[1]

If we allow the environment movement to become a pretty plaything of the affluent, and ignore the real environmental problems of the ghetto and the farm laborer and the blue-collar worker in his factory, then we will be choosing political suicide, and the name of our sword will be irrelevance.
Adam Walinsky, Earth Day—The Beginning[2]

About a year before I began work on this book I wrote an essay for the *World Policy Journal* entitled "American Environmentalism: A Movement Courting Irrelevance." I was motivated to say what I said then in response to the prophecies of Robert Nisbet and several other American scholars. Nisbet and his peers had predicted in various essays that American environmentalism would one day be seen as the most significant social movement of the twentieth century. I found their predictions challenging, if not preposterous.

How, I wondered, could the American environmental movement—which had been unable to meet so many of its own stated goals, had lost so much ground during the 1980s, and was now so devoid of vision—possibly accomplish enough in the remaining few years of the twentieth century to be considered more significant than the labor, women's, peace, civil, and human rights movements of earlier decades? Given the political tenor of the times, the condition of mainstream environmental organizations, and the intensity of the anti-environmental backlash in the United States, true significance before the end of the century seemed highly unlikely.

Of course, to be "significant" a social or political movement need not be triumphant or even finished. Social movements are, arguably, never finished. Yet one would expect considerably more progress than America's environmentalists have achieved in three decades of intensive, extremely well-funded activity.

It's not that environmentalists have completely failed. In fact, they have triumphed in many Earth-saving battles and fostered environmental sensibilities embraced to some degree by a vast majority of Americans. But in doing so they have been unable to produce a significant improvement in the country's environmental health. American land, air, and water are certainly in better shape than they would have been had the movement never existed, but they would be in far better condition had environmental leaders been bolder; more diverse in class, race, and gender; less compromising in battle; and less gentlemanly in their day-to-day dealings with adversaries. Over the past 30 years environmentalism has certainly risen close to the top of the American political agenda, but it has not prevailed as a movement, or as a paradigm.

Reading again the promises expressed in the preambles to the National Environmental Policy Act (NEPA), the Clean Air Act of 1970, and the Water Pollution Control Act of 1972, reveals the incipient failure of modern environmentalism. Those and other landmark bills crafted by federal politicians and environmentalists working together set out, among other things, to

- Halt environmental degradation before the end of the century
- Restore air, water, and soil quality to safe if not pure conditions in seven years
- Develop renewable energy systems
- Implement sustainable-yield forestry.

Scientific assessments of the environment made by the Council of Environmental Quality (even during the Reagan and Bush administrations), the Environmental Protection Agency (EPA), or any objective body show how far short of those goals we are. It is difficult to imagine how the environmental movement could achieve "most significant" status before the end of the millennium.

What I saw in American environmentalism when I began work on my essay were remnants of an organized conservation initiative that began a hundred years ago with great promise. That early initiative became the organizational infrastructure of a movement that broadened both its definition of environment and its agenda. After Earth Day 1970 it grew exponentially in numbers of followers, energy, skill, and financial resources. By Earth Day 1990 the American environmental movement had become a vast, incredibly wealthy complex of organizations dominated by a dozen or so large national groups centered, if not headquartered, in Washington, D.C. Together, at times in chorus, the "nationals" (as they came to be called) crafted an agenda and pursued a strategy based on the authority and good faith of the federal government.

Therein, I believe, lies the inherent weakness and vulnerability of the environmental movement. Civil authority and good faith regarding the environment have proven to be chimeras in Washington; they are real only in the imagination of environmental leaders. The hegemony of the nationals and their leaders, who had focussed their attention and resources on a federal strategy, prompted me to assert in the winter of 1991 that the American environmental movement was "dangerously courting irrelevance." The nationals had also made two other, near-fatal blunders: one was to alienate and undermine the grassroots of their own movement; the other was to misread and underestimate the fury of their antagonists.

The triumphs environmentalism can rightfully claim evoked bitter reactions that expressed themselves in the hostile presidential administration of Ronald Reagan and, ultimately, in a national anti-environmentalist backlash called the Wise Use movement. The environmental movement's response to Reagan's two-front war against the Environmental Protection Agency and Superfund was genteel and anemic. Direct mail copy writers effectively exploited the president's hostility to the environment. They created in Reagan and his cabinet a fundraising hydra that brought millions of dollars and hundreds of thousands of new members into the nationals. In fact, some social scientists credit Reagan with saving the movement from "natural decline."[3]

The mainstream movement responded to Reagan by forming the harmless and stubbornly elitist Group of 10 (later renamed the Green Group), creating its own irrelevance by remaining middle class and white, pursuing "designer issues" expedient to fundraising, focusing on Washington, lobbying the wrong committees, failing to move women and minorities into top jobs, building ephemeral memberships with direct mail, ignoring the voice of vast constituencies and, eventually—under the rubric of third-wave environmentalism—cozying up to America's worst environmental violators.

Considering all the blemishes I saw as I began to write this book, I was prepared to stay with my "courting irrelevance" subtext. But as I worked through my research I began to see the real importance of something that had happened in Washington in October of 1991. That event tempered my pessimism. It was the People of Color Environmental Leadership Summit. Although there were only about 300 delegates, it was a highly significant meeting, for it marked the first time so many nonwhite Americans who called themselves environmentalists had gathered in one place. African, Latin, Asian, Polynesian, and Native Americans from 50 states, Puerto Rico, the Marshall Islands, Central America, and Canada issued a call to arms for "environmental justice." The battle they initiated will, I believe, revitalize American environmentalism and could even transform it into the truly significant movement historians have predicted—perhaps not by the end of the century, but soon thereafter.

Although I believe that many organizations are still courting irrelevance, I changed the title of this book to leave room for optimism about the future of environmentalism in America. In my research, I found nothing to make me optimistic in the national organizations that once called themselves "The Group of 10." What prompted me to withdraw my discourteous subtitle was a whole new cadre of social activists, an environmental ad hocracy formed largely of people who, ironically enough, once derided environmentalists for their elitism and for drawing national attention away from what they perceived as more pressing social issues. The same people now proudly describe themselves as "enviros." The People of Color Convention signaled an expansion and transformation of environmentalism that will lead, I believe, to a truly American political *and* social movement.

Introduction

The Polite Revolution

Polite conservation leaves no marks . . . except scars upon the Earth that could have been avoided.

David Brower

Turn-of-the century environmentalists never used the words *environment*, or *environmentalist* to describe themselves or the natural world. Until the early 1960s they referred to themselves as *conservationists*, although more radical figures like John Muir insisted on the term *preservationist*. The word *environment* to describe an all-inclusive category comprised of both human and natural habitats did not come into common usage until Rachel Carson used it in her groundbreaking exposé *Silent Spring* in 1964. This is not to say that earlier conservationists were not environmentalists, but simply that they used different words to define their interests. After a brief foray into its origins, my history of the modern American environmental movement begins at the moment Rachel Carson first brought the word into the public vocabulary.

Most, but not all, American historians view conservationists and preservationists as the progenitors of modern environmentalism. A few interpret the emergence of environmentalism in the 1960s as a completely new phenomenon. When the designation *environmentalist* first appeared and the notion of protecting both natural and human environments entered the American consciousness, most conservationists were still preoccupied with preserving wilderness and protecting wildlife. Before long, however, new issues would be added to the environmental agenda.

Early conservationists were primarily prosperous white men, a mixture of hikers, campers, mountain climbers, hunters, and fishermen who looked to the wilderness as a place of recreation and refreshment from urban pursuits. Their organizations were clubby and exclusive; some admitted new members only if sponsored by existing members. The Sierra Club, for years a private mountaineering club, required two sponsors. Several groups had whites-only provisions in their original charters.

Wilderness preservation, a key objective of the early movement, meant protecting massive tracts of open land as hunting and camping grounds. Whether the purpose was fishing, hunting, or aesthetic enjoyment, the preserve was generally regarded as a place for a privileged few. Early national parks were often posted "whites only," and essays written by upper-class hunters like Robert Roosevelt (Theodore's uncle) extolled sportsmen who pursued game for pleasure not food. William Hornaday, director of the New York Zoological Society, appealed to its members to "take up their share of the white man's burden" by protecting the forests and mountains. Madison Grant, founder of the Zoological Society and of the Save the Redwoods League, condemned the hunting practices of "the inferior southern European races" and warned that "swarms of Polish Jews . . . and other worthless race types. . . . with their dwarf stature, peculiar mentality and ruthless concentration on self interest [were] being grafted upon the stock of the nation."[1] With supporters like Hornaday and Grant, it is small wonder that early American conservationists had difficulty creating a true social movement.

Even today, social historians remain undecided about whether American environmentalism can be defined as a social movement. It has labeled itself one frequently enough in its own literature. But is "social" really an appropriate adjective?

Over the years environmentalists have used many of the tactics of traditional social movements—political organizing, protest demonstrations, boycotts, nonviolent resistance, and so on. But were early American greens pushing for anything social beyond their exclusive reserves? Have contemporary "enviros" pursued any more socially radical aims

Acknowledgments

To thank everyone who contributed something to this book would lead to the sacrifice of many trees and slight the few without whom it could never have been written. So I'll mention the few and apologize to the many who go unacknowledged.

First I must mention Richard Caplan, former editor at the *World Policy Journal*, who called one day in mid-1991 inviting me to write an essay on American environmentalism. I froze in fear and named a dozen people who were better qualified. Richard persisted, saying he wanted a reporter who had no set or predictable position on the subject. He also offered to work with me closely; I have never had a more patient, collegial editor. Many of his insights and observations have been carried from the journal article into this book.

Next I must thank Edward "Teddy" Goldsmith, who showed the essay to Madeline Sunley at MIT Press and said, "You ought to have this fellow write something for you." Madeline rose to bait, assigned a book, and completely restored my wavering faith in book editors, as did my copy editor Roberta Clark, who made a manuscript out of a mess.

To Bill Moyers, whose encouraging letter and offer of financial support made it quite easy to raise the rest, I will always be grateful; as I will to Sophie and Derek Craighead, who provided not only financial support but also love, understanding, and a quiet place to work in the mountains of Montana.

Indispensable to a writer are other writers willing to be on call for consolation, commiseration, quick readings, and general bellyaching about sources, deadlines, paragraphs, or crises of confidence. Particular thanks in this category go to Doris Ober, Mark Hertsgaard, Michelle Syverson, Philip Fradkin, Mark Shapiro, Trin Yarborough, Jeff Gillenkirk, Sue Zakin, Joe Kane, Joanna Molloy, and particularly Catherine Caufield, without whom this book would be much longer.

Reliable and articulate sources are essential to the journalist's work. Some, of course, have been more forthcoming than others. The following people returned phone calls, granted interviews, and spent valuable time patiently explaining their work and views on whatever I asked about: John Adams, Robert Allen, Dana Alston, Carl Anthony, Richard Ayres, Mike Bader, Peter Bahouth, Dàn Becker, Jim Bensman, Peter Berg, Brent Blackwelder, Saul Bloom, Patty Brissendon, David Chatfield, Mike Clark, Will Collette, Barry Commoner, Hanna Creighton, Chris Desser, Bill Devall, Barbara Dudley, Sharon Duggan, Riley Dunlap, Brock Evans, Michael Fischer, David Foreman, Lois Gibbs, Teddy Goldsmith, Robert Gottlieb, Claire Greensfelder, Richard Grossman, Herb Gunther, Denis Hayes, Randy Hayes, David Helvarg, Tim Hermack, Ricard Hofrichter, Ted Howard, Josh Karliner, Andy Kerr, Martin Khor Kok Peng, Fred Krupp, Winona La Duke, Kathy Lerza, Brian Lipsett, Francesca Lyman, Andy Mahler (who started me thinking about civil authority), Jerry Mander, Eric Mann, Leah Margulies, Michael McCloskey, Carlos Melendrez, Al Meyerhoff, Margaret Morgan-Hubbard, John Moyers, Laura Nader, Gaylord Nelson, Bill Newsom, John Passacantando, Jane Perkins, Michelle Perrault, Joshua Reichert, Mark Ritchie, Bill Roberts, Jacob Scherr, Patty Schifferle, Karen Sheldon, Vic Sher, Gar Smith, Don Snow, Andy Szasz, James Thornton, Bill Turnage, Tom Turner, Kate van der Moire, Steve Viederman, Lori Wallach, and Margaret Hayes Young. Jay Hair agreed to talk to me for an hour, but only in person at his office, which is about 3,000 miles from mine. Unfortunately I had to pass up the offer. This book would have been far richer with his input.

I also want to thank some people I will probably never know: the three anonymous peer reviewers MIT Press selected to read the second to the last draft of the book. Although I tried to persuade the Press I had no peers (in vain, of course), they went out and found three people who clearly know far more about the subject than I do. Peer review is a nerve-racking process, especially the waiting, not knowing who the readers are but aware of their power to kill the book. Naturally I am pleased that all three readers recommended publication, but my deepest gratitude is for the time and effort they took to comment on my assumptions, logic, and arguments and correct my many errors of fact and spelling—none of them horrendous but all capable of diminishing the credibility of the work. So, whoever you are, thanks for wading through a messy and disorganized manuscript. Perhaps, one day, we will be able to debate face to face some of the points on which we clearly disagree.

Subsequent to publication of this book in early 1995, the 104th Congress of the United States demonstrated the frailty of a capital-centered national strategy. Gradual and suble abdication of federal authority over the environment became a thing of the past as Newt Gingrich's "Contract with America" accelerated the process.

In 1994 congressional Republican candidates gambled that environment was not as high on the American voters' agenda as environmental leaders had been telling them it was. Americans *were* voting green, exit polls indicated, but the green was cash and currency, not forage and foliage. The so-called "environmental vote" seemed shallow and ephemeral. "Economy first, ecology later," voters seemed to be saying. The gamble paid off.

Energized by victory, the first Republican Congress in forty years set out to challenge or eviscerate (depending on who was quoted) every environmental statute in the federal code and dismantle an environmental protection bureaucracy twenty-five years in the making. While it appeared, as we went to press with this edition, they may have overreached their bounds, begetting a tentative backlash—even within their own ranks—the fact remains that the mainstream environmental movement placed far too much faith in the federal government and failed to foster strong grassroots support for environmental protection.

The future of American environmentalism thus remains outside the Beltway, in the hands of a new civil authority forming in and around thousands of watersheds, forests, factories, and communities scattered throughout the country. I call it "The Fourth Wave."

of the environmentalist is a backpacker and a tree-hugger. Environmentalism means wildlife protection and wilderness conservation, while the environmental movement is identified with the Sierra Club and similar organizations. Only quite recently have clean up and prevention of toxic wastes, occupational health, and environmental justice became full partners in the American environmental agenda.

The nationals can fairly take credit for many of the century's environmental accomplishments, but they must also share blame for the troubling setbacks the environment has suffered. This is particularly true for the years of the Reagan and Bush administrations, when so many of the movement's earlier triumphs were eviscerated. During those years, when environmental conditions and political circumstances called for tougher, more confrontational tactics, most of the mainstream organizations lost the momentum they had developed in previous decades. In a desperate drive to win respectability and access in Washington, mainstream leaders politely pursued a course of accommodation and capitulation with elected officials, regulators, and polluters. Compromise, which had produced some limited gains for the movement in the 1970s, in the 1980s became the habitual response of the environmental establishment. It is still applied almost reflexively, even in the face of irreversible degradations. These compromises have pushed a once-effective movement to the brink of irrelevance.

Measuring the success or failure of a social or political movement is a subtle and difficult task involving the assessment of largely subjective, unquantifiable trends and developments. Is a movement successful when it has prevented things from becoming worse than they otherwise would be? Or should it be held accountable to the objectives stated in its own promotional and fundraising materials?

Dr. Barry Commoner, a biologist and lifelong environmental advocate, argues that it is easier to assess the impact of the environmental movement than to judge the success of other social movements because "the quality of the environment can be expressed in terms of generally unambiguous measurements." Since the beginning the early 1960s, he points out, sophisticated machines and devices have existed to calibrate the amounts of harmful pollutants in the air, soil, or water, in human body tissue, and in vast populations of wildlife. The readings from such devices are gathered, recorded, and analyzed by the Environmental Pro-

different strategy emerges. The significant statistic in this regard is that as membership in the nationals declined in the early 1990s total contributions to all environmental organizations was increasing. New groups, new concerns, and new strategies may still fulfill the prediction of some historians that American environmentalism will one day be regarded as the most significant social and political movement of the century.[6]

There are eight years left in the prophesied "environmental century." The overall environmental movement is vaster and more varied and ideologically diverse than it was on Earth Day 1. It still contains at its roots enough passion, determination, and vitality to achieve its aims. Unfortunately, however, political power has not reached those roots, nor has the real money. Both are still heavily concentrated in a few safe and respectable organizations that comprise what is known as "the mainstream."

The mainstream environmental movement is still dominated by a handful of the largest national organizations, all centered, if not headquartered in Washington, D.C. and heavily staffed by lawyers and MBAs. These organizations—most notably the Sierra Club, the National Wildlife Federation, the National Audubon Society, the Natural Resources Defense Council, the Wilderness Society, Friends of the Earth, the National Parks and Conservation Association, and the Environmental Defense Fund—have catalyzed and defined the American environmental effort over the past 25 years.

Since Earth Day 1, mainstream leaders have developed a self-confident conviction that their strategy—a legislative/litigative initiative focused largely on the federal government—is central to the environmental effort. They see regional grassroots activity, which receives little or no national media attention, as at best helpful, at worst an embarrassing sideshow to the main event. The main event, almost always, takes place in the national capital.

There is some justification for the mainstream's hubris. For 25 years, by sheer force of size and wealth, the organizations have defined environmentalism and set what the mass media perceives as the environmental agenda. The unfortunate result is that environmentalism has remained narrowly defined. In most people's minds it is equated with scenic nature and cuddly animals (mostly furry mammals). The image

in 1970 showed that 53 percent of Americans viewed "reduction of air and water pollution" as a national priority, up from 17 percent in 1965. By the 1990s well over 80 percent of Americans, according to recent polls, were comfortable calling themselves environmentalists. A recent *New York Times* poll found that 45 percent of Americans believe that "protecting the environment is so important that requirements and standards cannot be too high and continuing environmental improvements must be made regardless of cost."[3]

In 1980 Resources for the Future found that 7 percent of Americans (15 million people) regarded themselves as "environmentally active." A significant portion of them consider writing checks a form of activism, contributing hundreds of millions of dollars (close to half a billion in 1994) to a movement they hope will defend their environment. Never before have so many Americans supported a single cause so long or so generously.[4] American conservation philanthropists—whether two-figure or six-figure donors—may not be highly committed from an activist's standpoint, but their philanthropy has, until quite recently, been remarkably loyal.

In the late 1980s, however, two disturbing developments became evident. First, the large national organizations began to produce disappointing results with their supporters' millions. Interpretations as to why this happened vary widely, but it is worth noting that the decline in the organizations' potency and effectiveness coincided with their professionalization and a concurrent loss of passion within the ranks. Second, perhaps as a result of the nationals' loss of effectiveness, the constituency they had created, and then relied upon for sustenance, began to abandon them. As membership wanes in the 1990s, public concern for the environment is less frequently expressed through personal philanthropy toward the established organizations.[5]

In 1988, a Lou Harris poll found that 97 percent of Americans believed that more should be done to protect the environment; five years later, as small-donor contributions decreased sharply, the same question received an 82 percent response. It is possible to conclude that environmentalism is simply in "natural decline," a term used by social historians to describe the inevitable demise of all social movements. Or, alternatively, people may be expressing their environmental sentiments in different ways as a new environmental movement employing a very

than imploring people to consume responsibly and live ecologically? Despite the efforts of Robert Marshall, justice, the critical ingredient of any social movement, did not enter the agenda of American environmentalism until very recently.

The essential activism of environmentalism has thus differed significantly from other American social and political movements. It was not, like the labor, civil rights, or women's movements, a rebellion forged in oppression—although oppressed people did eventually come to see themselves as victims of environmental degradation. None of the Progressive founders or early adherents were enslaved, disadvantaged, dispossessed, or discriminated against. Quite the contrary. They were landed hunters and fishermen inspired by apolitical naturalists to protect the sources of their aesthetic pleasure, their game preserves, and their vision of America as the promised land of their Anglo-Saxon forefathers. By midcentury, although the conservation organizations were less exclusively enclaves of the very wealthy, their aims remained limited and conservative.

The first real sign that environmentalism had become politicized appeared on April 22, 1970, Earth Day 1, regarded by some historians as the temporal boundary between the conservation and environmental movements. The event heralded a national commitment to a healthier environment. Yet the very quiet tenor of the day, in contrast to the militancy of the antiwar movement, suggested that environmentalism would remain genteel, white, and very polite.

The social and economic elite who recognized a threat to their own interests in environment degradation joined the movement by writing checks to old-line conservation groups, penning letters to their congressional representatives, or signing a full-page ad in a newspaper. Some encouraged their children to study ecology and pursue careers in environmental law, biology, architecture, and economics. These solid members of the professional classes, who hesitated to join social or political movements that required direct contact with the dispossessed or had been condemned as "communist plots," had found a perfect home in the environmental movement.[2] It was respectable, safe, and polite. Today they are its leaders, still safe, still respectable, and still very polite.

Nonetheless, this new environmental leadership gradually mobilized millions of concerned, though mostly passive, adherents. A Gallup poll

tection Agency, other environmental agencies, and a host of academic laboratories. They provide a reliable assessment of environmental quality and a reasonable means of determining the impact of environmental activism.

There have been some notable improvements in such pollutants as airborne lead, mercury, polychlorinated biphenyls (PCBs), and strontium 90. By most measures of chemical emissions and residues in air, water, soil, and tissue, however, the bold legislative initiative launched in 1974 to improve the American environment has failed. While the environmental movement cannot be held solely accountable for that failure, it cannot fairly claim success either. The best that can be said with any degree of certainty is that if the movement had not existed things would be worse.

It is the thesis of this book that things could be much better and could still become so, but not without a very different environmental movement. I begin my argument in Chapter 1 by exploring the environmental imagination in its historical context—the uniquely American phenomenon generated from ideological seeds brought to the continent since the Ice Age by successive waves of settlers. "Without imagining justice there could be no justice" said civil rights leader Medgar Evers. The same is true of the environment. From its inception in all forms, the American environmental movement has been driven by imagination. Any movement that aspires to significance must stimulate and foster in the minds of its followers and activists a vivid image of a better world. If we cannot imagine a healthy, bountiful, and sustaining environment today, it will elude us tomorrow.

In Chapter 2 I discuss the culture of reform that took hold of the movement in the 1960s and persisted until quite recently, when poorer Americans of all colors and ethnicities suddenly realized that they were victims of environmental injustice. In Chapter 3, fix becomes folly as the impact of counter-environmentalism reaches a historical climax during the administration of Ronald Reagan. The Reagan years, I postulate, were the decisive decade for the mainstream environmental movement. Instead of going *mano a mano* with the most environmentally hostile president in recent history, the movement blinked. In Chapter 4, I recount how mainstream organizations allowed a host of antagonists to gain influence virtually unchallenged, raised a lot of money, and invented something called "the third wave." In Chapter 5, I describe this conciliatory, amoral branch of reform environmentalism co-opted and confounded by coercive harmony.

In Chapter 6 I discern signs of a rejuvenated, angry, and decidedly impolite movement for environmental justice as people of color and working people who once shunned a movement they believed was threatening their security, begin to join up and, ever so reluctantly, call themselves environmentalists. Fighting for a healthier environment on civil rights grounds will, I argue in this chapter, redirect environmental litigation, strengthen the legal case against toxic pollution, and broaden the environmental constituency. In Chapter 7 I pause to assess the state of a faded green movement as it attempts, one last time, to pursue a federal strategy with an allegedly friendly administration. Finally, in Chapter 8, the soul of the book, I describe a new era of American environmentalism—in part, I admit, a figment of my own hopeful environmental imagination.

American environmental history to date can, I believe, be divided into three waves. The first began with the conservationist/preservationist impulse of the late nineteenth and early twentieth centuries and coincided with the closing of the frontier. The second wave came in the brief era of environmental legislation that began in the mid-1960s and was abruptly halted by the Reagan administration in the 1980s. The third wave, a relatively fruitless and hopefully brief attempt to find a harmonious ("win-win") conciliation between conservative environmentalists and corporate polluters, is with us as we approach the mid-1990s.

I believe, however, that a fourth wave is forming and that before the end of the century it will become the heart of a new American environmental movement. As it builds, the polite, ineffectual white gentleman's club that defined American environmentalism for almost a hundred years will either shrink into historical irrelevance or become an effective but equal player in the new movement. After so many decades of polite activism, the movement is becoming appropriately rude and decidedly American. If it becomes truly American by adopting new, wider strategies and a democratic ethos, it will prevail. I believe that will happen. If I am right, the *real* environmental movement has barely begun.

1

The Environmental Imagination

The force that through the green fuse drives the flower drives my green age.
Dylan Thomas

Almost every civilization in history has produced philosophers, poets, and prophets who imagined living in harmony with nature in a pristine, healthy environment. Some, mostly "primitive" societies realized their vision, at least for a while. Others have only been able to imagine it. Yet imagination is a powerful force. It can lead to reformation, enlightenment, or revolution. Or it can foster a political movement. America is the first civilization in history to turn its environmental imagination into a political movement.

The antecedents of America's environmental imagination are deep, varied, and often contradictory. The settling of the North American continent lasted for thousands of years as successive waves of migrants brought with them the environmental traditions of many lands, cultures, and religions. Gradually, the synthesis of their ideas created a unique and powerful vision of a world both prosperous and healthy, dynamic and sustainable, a world that, sadly, may never exist beyond our imagination, in spite of concerted national efforts to realize it.

If consciousness were the final measure of a significant social movement, American environmentalism could boast of great import, even of triumph, for Americans love their mountains, prairies, forests, rivers and meadows. And, by and large, they are aware that the environment is in jeopardy. Many tell pollsters they are willing to sacrifice a little of other parts of the American dream to protect natural bounties. Nonetheless, after more than a hundred years of intensive activism, human life in harmony with nature and a healthy environment still exist only in our imagination. The conservation movement, now over a hundred years old, and the environmental movement, now almost thirty, have yet to prevail.

Seeds

Wilderness remains because we allow it to exist.
Roderick Frazier Nash

There is no general theory of political movements. Once begun, they tend to be shaped by their historical and political contexts, taking on the characteristics of the society they are trying to change, acquiring the customs, organizational structures, traditions, styles of leadership, and divisions of labor found in the surrounding culture. Democracy necessitates very different strategies than bureaucratic socialism or monarchy. Thus a movement formed in twentieth-century America will be quite different from one that took shape in eighteenth-century Europe or nineteenth-century Russia, even if the goals and agenda are the same, and even if the origins of the contemporary movement come from the philosophies and social movements of the past.

The seeds of American environmentalism were sown long before European settlers arrived to challenge the land, and long before an environmental movement was organized. They were sown in Europe by philosophers and writers who watched the gradual degradation of their own continent, and in "Turtle Island" itself by the pre-European settlers who attained ecological wisdom by living close to the land.

Early white settlers carried with them to the "New World" some version of the Judaeo-Christian Bible. Many of them read little else. In spite of the hardships of the early settlement period, many pilgrims believed they had rediscovered the garden of Eden, the paradise lost in the old country and refound on this vast continent teeming with abundance.[1] While most heeded God's commandment to subdue the earth, a few settlers also remembered the earlier command to replenish it.[2] They became America's earliest conservationists. The continent's earliest *ecologists,* of course, had been here for thousands of years.

Some historians and essayists place blame for the state of the nation's environmental health on the Judaeo-Christian tradition. They see in it worship of a creative deity who is separate from its creation and who prophesies an apocalypse that requires destruction of the existing environment before a new heaven on Earth can be born. Though few such critics are surprised to find self-described apocalyptics like James Watt and Ronald Reagan among the adversaries of environmentalism, they are baffled by the fact that most of the environmental movement's founders and leaders have also been Christians. Some, like John Muir, came from fundamentalist-apocalyptic backgrounds. Indeed, it was precisely the Judaeo-Christian concepts of land ownership and stewardship that created the first environmental conflict on the American continents.

Earlier settlers, who had occupied the Americas for so long they eventually became known as Native Americans, had regarded the land as a commons for at least ten thousand years. To them, stewardship of the commons was an assumed tenet of the social contract, not something that needed to be debated, preached, or taught in school. The 100 million or so hoofed animals that grazed the commons—the bison, deer, elk, caribou, antelope, wild sheep, goats, and boar—were the peoples' livestock, to be culled when needed for food and clothing. The desire to eliminate most of them and replace them with hoofed domestic animals kept in fenced pastures and corrals seemed strange to Native Americans, whose idea of chattel was limited to beads and pottery. But that was how agricultural husbandry and property was being organized in Europe, and white settlers saw no compelling reason to change their farming methods or the land ethic that went with it. The commons was enclosed, and the seeds of environmental tension were sown in the New World by the notion of deeded land, fencing, and private property.

Nonetheless, the wisdom of those who lived here before survives. Their descendants often remind us that most of what Euro-American preservationists and conservationists have fought to preserve or conserve was actually stolen from their people. Archeologists have shown that what we call "wilderness" was not "unoccupied," as so often claimed, "since the last Ice Age." Nor, as early American philosophers and, later, direct mail copywriters have asserted, was it ever "frozen in time." For thousands of years most of the continent was a fecund hunting and fishing ground for peoples who only recently were expelled from it, as the Shosone, Bannock, Blackfoot, and Crow were forced to leave what is now called Yellowstone Park in 1872.

Many of the environmental sentiments we attribute to Euro-American nature philosophers like Henry David Thoreau, Aldo Leopold, and John Muir, and enlightened ecologists like Rachel Carson were embraced for millennia, but never written down, by these former occupants of Yellowstone Park. Their influence on the world's environmental imagination is still powerful. Many of the "new" environmental ideologies—deep ecology, sustainable development, and bioregionalism—are rooted in the unwritten ethics of Native American wisdom-keepers, whose stewardship of earth and concern for the seventh generation was only briefly stifled by European settlement. Most contemporary environmentalists, however, look to their own roots for wisdom, and the resultant cultural divide between Native and European Americans remains wide.

Drawing from Plato's *Republic,* Jean-Jacques Rousseau lamented the corruption of nature in humanity he believed was induced by property, agriculture, technology, and commerce.[3] Like Sir Thomas More in *Utopia,* Rousseau was critical of existing mores and values and sought a design to reconstruct society. In his *Discourses* he challenged the belief that better technologies, material wealth, and knowledge would lead to the improvement of humanity and morality. Large commercial centers, he warned, were bad for the human spirit. He prescribed instead the formation of cooperative agrarian communities.

With the exception of a few small bands of utopians who settled briefly in the Midwest, European settlers and early Americans did not heed Rousseau. Most preferred the ruminations of scientists like Francis Bacon and Isaac Newton who, along with Enlightenment philosophers René Descartes, David Hume, and John Locke, created a world view that desacralized nature and provided ideological fuel for the industrial revolution.

Bacon referred to nature as "a common harlot," which he hoped mankind would "conquer and subdue . . . and shake to her foundations." Descartes invoked Bacon in his defense of vivisection: animals, he said, were "soulless automata" whose screams under torture were "the mere clatter of gears and mechanisms." To Newton the world was a clock, wound by God: "The entrepreneur, merchant, industrialist scientist [were] God's counterparts, the skilled technicians that used the same mechanical laws and principles that operated in the universe to assemble the stuff of nature and set in motion the industrial production of the modern age." Locke created the anti-ecological creed that justified the commercial exploitation of natural resources: "Land that is left wholly to nature," he wrote, " is called, as indeed it is, waste."[4]

In his retreat from civilization, however, one American Transcendentalist drew upon his reading of Rousseau, and from Thoreau's search for "the tonic of wildness" came the preservationist seed of the American environmental imagination. Although preservationism would come to clash with later ideas of conservation and the "wise use" of wilderness, it would persist in the environmental imagination as a driving force of the movement through the end of the twentieth century.

Thoreau's concern for wildlife was atypical of the rapidly expanding industrial north: "Who hears the fishes when they cry?" he asked the lowly shad, whose migratory route was interrupted by dams on the Concord and Merrimack rivers. The wilderness buckaroos who a few generations hence formed Earth First! could have as easily looked to Thoreau for inspiration as to their anarchist mentor Edward Abbey, for Thoreau had definite eco-saboteur leanings. "I for one am with thee," he wrote to the shad, "and who knows what may avail a crowbar against the Billerica Dam?" The author of "Civil Disobedience," of course, served time in prison for defying the tax collector.

Thoreau also contributed to the American environmental imagination the notion that the biosphere is a unit capable of certain limited mechanisms of self-protection. "The earth I tread on is not a dead, inert mass," Thoreau wrote, "it is a body, it has a spirit, is organic and fluid to the influence of its spirit."[5] The idea itself was already almost a century old. In 1785 James Hutton, the "father of geology," delivered a lecture to the Royal Society of Edinburgh, in which he propounded a theory that the earth was alive. That such a subversive thought could survive two centuries of technological culture and become a seed of contemporary environmental thought is itself amazing. But it did. In 1969, more than a century after Thoreau's death, an English scientist named James Lovelock gave the image of living earth a name. At a Princeton University meeting about the origins of life Lovelock introduced "the Gaia Hypothesis." Gaia, although still somewhat marginalized in New Age culture, is now seriously debated by American scientists in almost every discipline and has become integral to ecological discourse.[6]

Wilderness WASPs

Many of the earliest settlers of "the new world" sought to escape the limitations imposed on them by economic and religious elites. They were looking for virgin land, which many proceeded to ravage, much as dynastic and civil wars had periodically devastated the European landscape. Although most white settlers were hardly environmentalists in the contemporary sense of the word, among their descendants were a few with conservationist impulses. People like Daniel Boone, John James Audubon, and George Bird Grinnell saw in the grandeur of North America many things and places worth protecting from the growing environmental degradation of industrial Europe. Gradually naturalist writers and philosophers began advocating stewardship of the land and the preservation of nature for its own sake. They wrote and taught at a time when the despoliation of the American continent had been under way for well over 200 years.

During the seventeenth century the entire beaver population of New England was wiped out by Anglo-Dutch fur traders. In the eighteenth century Ben Franklin hoped for a population large enough to "cover the whole northern if not southern continent." Vast forests were cut down to create farms, towns, and, later, to build railroads and factories. In the east rapid population growth, the development of steam power, and the seeming inexhaustibility of the continent's natural resources would soon produce squalid urban ghettos and noisome manufacturing towns. On the Great Plains bison were slaughtered by the million and left to rot.[7] The frontier mentality and an early version of Manifest Destiny combined to create a wasteful, rapacious culture opposed by few and encouraged by our most revered forefathers. The only people in the way of such developments were driven from arable land onto unproductive reservations from which they watched their conquerors destroy what had been, for as long they could remember, their garden.

A few Euro-Americans began to notice that forests no longer seemed to stretch into infinity, passenger pigeons didn't block out the sun for an hour at a time, and rivers weren't surging with fish as they once had been. The wanton waste of resources and the compulsion to use them for short-term gains were called to task. One of the first to do so was John Wesley Powell, a geologist and explorer who believed that the western part of the continent, with its vast deserts and mountainous regions, should be settled and managed very differently than the east. Powell's elaborate prescription for smaller homesteads, irrigation districts, and official designation of areas for mining, grazing, and forestry could be seen as America's first "sustainable development" plan. Unfortunately, Powell was ignored by all but a handful of fellow visionaries.

A decade later, however, America's few conservationists began to organize. In 1892 they organized the Sierra Club, and in 1896 National Audubon chapters opened in New York and Massachusetts. Members of the new organizations were mostly well-bred hunters, fishermen, and campers.[8] Virtually all were white Anglo-Saxons, an ethnic group that would dominate American environmentalism until the late twentieth century. The gospel of the hour was efficiency, the management of resources the new science. Equity, the notion that natural resources belonged to all people, was debated, eventually accepted, and written into laws whose interpretations are still being challenged in court. And Gifford Pinchot made a name for himself.

Pinchot, a wealthy young traveler, was trained in forestry and committed to public service. In Europe he saw the consequences of rampant resource exploitation and studied the German art of forest management. Returning to America around the turn of the century, he noticed for the first time the "flurry of development" at home. "The American Colossus," he later wrote, "was fiercely intent on appropriating and expropriating the riches of the richest of all continents—grasping with both hands, reaping where he had not sown, wasting what he thought would last forever. New railroads were opening new territory. The exploiters were pushing farther and farther into the wilderness."[9]

Pinchot was a member in good standing of the Boone and Crockett Club, an elite hunting association. There he met and became close friends with an up-and-coming young Republican politician named Theodore Roosevelt. Later President William McKinley appointed Pinchot to head the Agriculture Department's Forestry Division. In 1901 when McKinley was assassinated, the new president remembered his conversations with the young forester at Boone and Crockett.

Pinchot had observed that America's natural resources were being exploited at an alarming rate by businessmen with little or no concern for the future of the land. He persuaded Roosevelt that the appetites of many of his own largest supporters needed to be contained. In so doing Pinchot coined a term that would be coopted by a future generation of land barons, miners, and timber companies—"wise use." Assuring the wise use of resources, Pinchot argued, required a federal bureau to oversee public lands and protect them from unwise use. Moreover, he informed Roosevelt, the Department of Agriculture had only a few forests under its control, so a new agency was needed to protect the nation's vast woodlands.

In 1905 the U.S. Forest Service was established in the Department of Agriculture. In time, the service would become the obsequious servant of the timber industry. But not under Pinchot's leadership. In the context of America's spasm of land expropriation, the idea of a national forest service seemed an essential part of an overall program of Progressive reform.

Countering Pinchot's utilitarianism was a feisty Scottish-American named John Muir, who had moved west to California as a young man and discovered the magnificent Sierra Nevada mountains. Muir, a strict Presbyterian turned nature Romantic, spent weeks alone hiking from peak to peak through the white granite splendor of the high Sierras. There he developed the strong preservationist sentiments that would evolve into the second important strand of the early environmental movement. Muir believed deeply in the redemptive powers of nature and argued in his writing and speaking that wilderness deserved to exist for its own sake. While he agreed with Pinchot that careful Prussian-style forestry needed to be applied in some parts of the country, it was also, he insisted, essential to protect certain areas, whole eco-systems in fact, from *all* resource exploitation. They must be preserved forever as virgin, roadless, damless, and mineless, accessible to humanity only on foot, and only to those willing to sleep on the ground and carry out everything they carried in. Muir—who founded the Sierra Club in 1892 in partnership with Robert Underwood Johnson—remains today at the heart of the American environmental imagination, which has yet to actualize his vision.

Muir also lobbied Roosevelt. They met in Yosemite Valley in the spring of 1903. As they stood on the valley floor, surrounded with wildflowers and steep granite walls, Muir told the president that everything he saw around him should be protected by legislation as a cathedral where people could experience the wild in their souls and rejuvenate themselves after months in the urban jungles of America. Roosevelt liked the sentiment and, a few months later, introduced a bill to protect Yosemite Valley and its surrounding peaks, already a national park, from private exploitation and environmental damage. The formation of national parks on federal land became a legacy of the first Roosevelt administration.

Muir and Pinchot presented to the Roosevelt White House, and through it the entire nation, conflicting environmental visions—conservation (Pinchot) versus preservation (Muir). Their differences can best be illustrated by their conflicting positions on the 1912 damming of the Hetch Hetchy Valley, a project designed to provide fresh water to the city of San Francisco over 200 miles away. "These temple destroyers, devotées of ravaging commercialism, seem to have perfect contempt for Nature," wrote Muir. "Instead of lifting their eyes to the God of the Mountains, they lift them to the Almighty Dollar." Of the same project Pinchot said: "I am fully persuaded that by substituting a lake for the present swampy floor of the valley, the injury is altogether unimportant compared to the benefits to be derived from its uses as a reservoir."[10] After a bitter fight Hetch Hetchy was dammed.

Conservationism and preservationism thus became the primary antecedents of American environmentalism, and the collision that occurred between them then persists to this day. Pinchot, who eventually became director of the U.S. Forest Service, convened huge conferences of state governors and federal officials to discuss and plan the wise use and conservation of natural resources. He deliberately excluded Muir from either the program or the invitation list. Muir, for his part, devoted his writing, speaking, and organizing skills to building the foundation of what would become known, ironically, as the conservation movement. Meanwhile "enlightened" leaders in the corporate sector gradually accepted, even embraced Pinchot's conservation ideas. Eventually resource conservation and wise use became the dominant paradigm of modern American environmentalism.

Pinchot himself could hardly have foreseen that the bureaucracy he thought would contain the greed and rapaciousness of timber men and miners would fall victim to the influence and, at times, corruption of the very people it was designed to contain. His failure of foresight and his faith in bureaucracy shaped American environmentalism by creating the need for a movement to serve as watchdog to the agencies he had inspired.

Land Ethos

We still profess our love of the land and its treasures, but our love rarely interferes with our abuse of it.

Philip Shabecoff, Green Wire

Although Mormons had practiced land stewardship and accepted the concept of flowing water as common property—an ethic they learned from their neighbors, the Mountain Utes of southern Utah—not until 1949 did an articulate land ethic appear in American literature.[11] That was the year Aldo Leopold published his *Sand County Almanac*. Leopold, a pioneering ecologist who worked under Pinchot at the U.S. Forest Service, was among the earliest government officials to see things John Muir's way and support setting aside vast expanses of land as wilderness. Like Muir, he believed that virgin nature has value in its own right.

Leopold came to his ecological world view gradually, by way of the ideological routes mapped out by conservationist predecessors. But as he grew older, the preservationist arguments made more and more sense to him. When he worked under Pinchot he accepted scientific forestry, game management, and the government agencies formed to direct them. Then one day in 1909, on a predator-control hunt in mountainous New Mexico, he shot a mother wolf. As he approached her dying frame he saw in her eyes "a fierce green fire." "I realized then and have known ever since, that there was something new to me in those eyes—something known only to her and to the mountain. I was young then, and full of trigger itch; I thought that because fewer wolves meant more deer, no wolves meant a hunter's paradise. But after seeing the green fire die, I sensed that neither the wolf nor the mountain agreed with such a view." [12] "Green fire" became a common image in the American environmental imagination, and "thinking like a mountain" a way of seeing the world.

In the *Almanac* written almost 40 years after the killing of the mother wolf, Leopold repeats Thoreau's adage that "in wildness is the preservation of the world" but expands the concept of preservation beyond the spiritual to the biological. The wilderness, he asserts, is the world's only storehouse of genetic diversity, without which we will all perish. He called for a "land-based ecology" and a new ethic that "simply enlarges the boundaries of community to include soils, waters, plants, and animals." Leopold's land ethic "changes the role of *homo sapiens* from conqueror of the land-community to plain member and citizen of it. . . . All ethics," Leopold declared, "rest upon a single premise: that the individual is a member of a community of interdependent parts. . . . A thing is right when it tends to preserve the integrity and stability and beauty of the biotic community. It is wrong when it tends otherwise."[13] These sentiments and his insistence that vast sections of open land be deeded to the commons brought outcries and even occasional red-baiting from colleagues. Some environmentalists, like Jay "Ding" Darling, founder of the National Wildlife Federation (1935), opposed Leopold's ideas, fearing they would "lead to the socialization of property."[14]

Darling's fears notwithstanding, over 40 percent of America's land has come into public domain, largely due to the efforts of conservationists. One environmental historian, Donald Worster, credits them with saving the realm. "Indeed," writes Worster, "one of the most effective ways our democracy has devised to rescue itself from near extinction at the hands of holders of private wealth has been through public land ownership. That discovery has been part of the legacy of the American conservation movement."[15]

Tocsins and Toxins

For the first time in the 3.5-billion-year history of life on this planet, living things are burdened with a host of manmade poisonous substances, the vast majority of which are now even more prevalent in animal tissue and the elements than they were twenty years ago when Earth Day first imposed itself on the popular consciousness.

Barry Commoner, Making Peace with the Planet

American conservationists and preservationists both promoted their cause and primed the environmental imagination with seductive images of nature. In the nineteenth century the immense canvasses of Albert Bierstadt and Thomas Cole of the Hudson River School and wildlife paintings by naturalist John James Audubon fed the American imagination with inspiring images of the wilds. In the twentieth century, Ansel Adams, Elliot Porter, Phil Hyde, and Edward Weston used the camera to bring conservation aesthetics to a vast following of urban Romantics, many of whom had never set foot in the woods or slept on the ground but were awed by photographs of the Tetons at sunset and Yosemite by moonlight.[16]

While magnificent art and photography helped sustain the impulse to preserve the wilderness, it had the unfortunate result of defining environment as an ex-urban phenomenon separate from most people's daily lives. From a graphic standpoint, the urban environment was a lost cause. Next to arresting photographs of Yosemite or the Grand Canyon, the muckrakers' depiction of the inner city environmental story was a hard sell indeed. It still is. Lacking aesthetic images, it was many decades before urban environmentalists began to seek public support by invoking the injustice of their environment.

In part because of the influence of aesthetics, for almost 100 years saving the nonhuman domains of Earth was the single motive of American environmentalism. Until very recently that objective remained unattached to all other social movements, even those—like public health and city parks—that had clear environmental dimensions. American environmentalists rarely questioned or challenged the technologies that allowed them to communicate, travel, publish, or promote their cause—technologies with considerable environmental impact. Thus the effects of toxic pollution or improvements to human health and occupational safety—important adjuncts to other social movements—were relative latecomers to environmental thinking.

In 1962, toxic compounds and their impact on the biota were first documented for public consumption in *Silent Spring,* Rachel Carson's masterful account of pesticide and other chemical pollutions. "The most alarming of all man's assaults upon the environment," Carson wrote, "is the contamination of air, earth, rivers and sea with dangerous and even lethal materials. . . . For the first time in the history of the world, every human being is now subjected to contact with dangerous chemicals, from the moment of conception until death. . . . The question is whether any civilization can wage such relentless war on life without destroying itself, and without losing the right to be called civilized." Those sentences not only sounded the alarm bell that sparked a rebellious protest movement, they popularized a word few Americans had ever seen, heard, or used in quite that context—the *environment.* [17]

Although Rachel Carson will survive in history as the godmother of modern environmentalism, urban planner Robert Gottlieb, in his landmark revisionist history of environmentalism, *Forcing the Spring,* resurrects an earlier, unheralded seed of anti-toxics environmentalism. Alice Hamilton, "the mother of American occupational and community health," as Gottlieb describes her, was "this country's first great urban industrial environmentalist."

While preparing for medical studies in Michigan during the late 1880s, Hamilton became a social reformer. Later, as a lecturer in pathology at the Women's Medical School of Northwestern University, she became active at the Hull House settlement in Chicago. Jane Addams and other reformers were crusading for improved sewage systems, garbage collection, and clean water, and against typhoid, carbon monoxide pollution, tetraethyl lead, and horse manure on city streets (a troubling effluent of the time).[18] Between its opening in 1889 and Addams' death in 1935 Hull House was a meeting place for social activists and humanitarians throughout the Midwest. There Hamilton met union organizers who regaled her with graphic stories about the unhealthfulness and dangers of the American industrial workplace—particularly exposures to white phosphorus and lead. Hamilton shifted her efforts from medicine to a crusade for occupational health and safety.

She came to believe that American workers were used as "laboratory material" for manufacturers, particularly industrial chemists who were exposing their workers to solvents and other toxins without adequate ventilation or protection. In 1919 she was appointed assistant professor of industrial medicine at Harvard University, shortly after which she became a early critic of leaded gasoline, which she believed could never be made safe and should be banned from use in automobiles. "Where there is lead, some case of lead poisoning sooner or later develops," she wrote presciently in 1925. After publication of her groundbreaking *Industrial Poisons in the United States,* Alice Hamilton became a recognized expert in urban environmental health, long before the word "environmental" came into common use.[19]

Unfortunately neither her sensibilities nor her activism caught hold of the nation's environmental imagination. Occupational health became a sidebar to the labor movement and urban pollution a weak sister of public health. It would not be until long after her death that Alice Hamilton's perspective would join the agenda of American environmentalism, and it would be another two decades before a sensitive historian would see her for what she was—an environmentalist.

During the 1970s and 1980s public and occupational health and safety finally joined the movement for wilderness protection as a critical concern of environmentalism. And in the early 1990s, when the federal government's newly implemented Toxics Release Inventory revealed that American companies actually admitted to releasing more than five billion pounds of over 300 highly toxic compounds into the environment every year, the national movement against toxic pollution grew from a minor subsidiary of mainstream environmentalism to a powerful national initiative. It was fueled by essentially the same impulse that had inspired the conservation movement of earlier decades—the vision of a safe and healthy environment for the biotic community—now enlarged to include human Americans wherever they lived.

Earth Days

Earth Day was not a spontaneous uprising.
Philip Shabecoff, Green Wire

Every day is Earth Day.
Overused slogan of Earth Day 1990

Historians begin their account of modern American environmentalism at various points and with various events and episodes.[20] Earth Day 1970 is the most common and least imaginative of the starting points. It was an historic and momentous day, but it was not the beginning of a movement. Angela Mertig and Riley Dunlap believe that the conservation movement became the environmental movement in 1967, the founding year of the Environmental Defense Fund, "the first of a new breed of national environmental organizations."[21] The founding of EDF too was an important milestone, but the metamorphosis from conservation politics to environmentalism was already under way in 1967.

The best starting point for modern environmentalism, if there must be one, is probably 1962, the publication year of *Silent Spring*. Carson's book—along with Paul Ehrlich's *Population Bomb,* a neo-Malthusian tract on human population, and Barry Commoner's *Closing Circle,* which rephrased ecological verities like "everything is connected to everything else" and "everything must go somewhere"—alarmed, angered, and aroused a broad new constituency of middle-class activists. Together with traditional conservationists and a few enlightened politicians they began to draft federal legislation: the Wilderness Act in 1964, the first Clean Water Act in 1965, the Clean Air Act (1967), and the Wild and Scenic Rivers Act (1968), all of which passed with healthy majorities. During those years the word *environment,* with all its assorted variants, entered the American vocabulary as a term that accrued more and more meaning as our sense of place in the universe and our understanding of ecology expanded together. Environmentalism, which by then had absorbed the conservation movement, seemed here to stay.

As we have seen, the agenda of the American environmental movement did not emerge ready made during the 1960s. A vast tradition existed before Carson set pen to paper, under several different names, none of which included the *e*-word. What became known as environmentalism was an amalgam of resource conservation, wilderness preservation, public health reform, population control, ecology, energy conservation, anti-pollution regulation, and occupational health campaigns. All had become separate public concerns and developed as self-contained mini-movements during previous decades and, even, centuries. The modern environmental movement evolved from these many issues and causes in the context of a post–World-War-II urban environment whose degradation had become insistently obvious to people of all classes and races.

By the end of the 1960s, therefore, the momentum of the new movement was well under way. Earth Day 1970 was a culmination, not a beginning. Or, as environmental historian Samuel Hays puts it, "Earth Day was as much a result as a cause. It came after a decade or more of evolution in attitudes and action without which it would not have been possible."[22] Hays lists Earth Day 1970, together with environmental "disasters" such as the ignition of the Cuyahoga River and the Santa Barbara oil spill, as episodes that advanced the environmental initiative in America toward passage of what most historians consider the ultimate landmark of the new movement, the National Environmental Policy Act of 1970. (For more on NEPA see Chapter 3.)

Earth Day 1 was a coordinated nationwide outpouring of environmental sentiment celebrated by about 20 million Americans nationwide. People of every occupation and persuasion stopped what they were doing for a few hours to express their concern for something they loosely defined as their "environment." To some it meant wilderness, to others it meant their backyard. To all of them it meant their health—if not physical, spiritual. Whatever the definition, Earth Day 1 was the largest one-day outpouring of public support for any social cause in American history.

The theme for the day was set in Washington D.C. by Denis Hayes, a 25-year-old Harvard Law student who told a national rally that "Our country is stealing from poorer nations and from generations yet unborn. . . . We're tired of being told we are to blame for corporate depredations . . . institutions have no conscience. If we want them to do what is right, we must make them do what is right."[23]

Reviews of the day from the right and left were mixed. Journalist and social commentator I. F. Stone called it a "gigantic snowjob" that diverted attention from the widening war in Indochina. "Just as the Caesars once used bread and circuses," he wrote, "so ours use rock and roll, idealism and non-inflammatory social issues to turn the youth off from more urgent concerns which might really threaten our power structure." *Ramparts* magazine, a leading voice of the left, agreed, calling Earth Day "the first step in a con game that will do little more than abuse the environment even further." He didn't say it on Earth Day, but the sentiments of African Americans at the time were represented by the belief of Urban League President Whitney Young that "the war on pollution is one that should be waged after the war on poverty is won."

Earth Day's creator and Chairman Senator Gaylord Nelson (D-WI) didn't help relations with other American movements by asserting that "the most critical issue facing mankind" was the environmental crisis, which made "Vietnam, nuclear war, hunger, decaying cities, and all other major problems one could name . . . relatively insignificant by comparison."[24] Nor did Denis Hayes ingratiate himself to other activists. He was determined to "involve the whole society" in the day's events. Although he said, "Our goal is not to clean the air while leaving slums and ghettos, nor is it to provide a healthy world for racial oppression and war," many antiwar and civil rights leaders feared Earth Day would deflect interest from their issues.

Mainstream environmental organizations' responses to Earth Day were mixed. The *ancien régime* represented by the National Wildlife Federation, National Audubon, and the Sierra Club were wary of participating for fear attention would be drawn away from traditional conservation issues. "We cannot let up on the battles for old-fashioned wilderness areas," cautioned Sierra Club Vice President Ed Wayburn.[25] If the record is complete, no Earth Day speakers said he had to.

Stewart Brandborg, the last true activist to lead the Wilderness Society, sympathized with Earth Day and provided money and office space for the project. More surprising was the support of the Conservation Foundation, which provided a $20,000 loan to Hayes and Nelson when they were desperate for cash. At the time the foundation was run by Sydney Howe, one of the first mainstream leaders to take a genuine interest in urban environmental issues, which he saw as inseparable from civil rights and other social justice movements. (Howe was fired by the CF board in 1973.)

There have been celebrations of Earth Day on April 22 every year since 1970, with larger events occurring at the decades. Semi-permanent Earth Day committees have been formed in almost every major American city to coordinate relevant events, demonstrations, teach-ins, and concerts. Environmental Action, the national organization formed to coordinate Earth Day 1970, has permanent offices in Washington with a full-time staff of lobbyists, researchers, and educators. The group is best known for its annual announcement of "the Dirty Dozen"—companies with the worst pollution records in the country.

The tenth anniversary celebration of Earth Day planned in 1980 fizzled out when it was revealed the main event was being run by Byron Kennard of Environmentalists for Carter. Half of the $200,000 budget was provided by the Carter administration, which also allowed organizers to use telephones and offices at the Department of Energy. Many activists perceived the whole endeavor, probably correctly, as an adjunct to the Carter re-election campaign.

Between Earth Day 1970 and Earth Day 1990, the next large-scale celebration, the world became more toxic. Even Denis Hayes, who by 1990 had become considerably more conservative than the young radical who ran the first event, could see it. "By any number of criteria you can apply to the sustainability of the planet," he said, "we are in vastly worse shape than we were in 1970, despite twenty years of effort."[26]

Things would certainly have been even worse had the environmental movement not expanded so rapidly. In the years between Earth Days issues arose about as fast as species became extinct. The environmental agenda expanded and new organizations sprouted around each new crisis. The nationals became large and professionalized, and environmental protection became the law of the land. But given the state of America's land, air, and water there wasn't much to celebrate. The very definition of "environmentalism" was up for grabs. And so was Earth Day.

Denis Hayes again headed the national coordinating group in 1990, but this time he had a $3-million budget, a blue ribbon board of advisors, a staff of hundreds, and offices in almost 20 cities. George Bush, the Chemical Manufacturers Association, and about 80 percent of the American population had declared themselves green. Corporations like Monsanto, the manufacturer of PCBs and herbicides and the developer of saccharin and recombinant bovine growth hormone, offered to sponsor Earth Day activities in several cities. (The St. Louis Earth Day committee accepted $15,000.) British Petroleum donated $7,500 to cash-starved Earth Day festivities in Cleveland—a small price for a quick citywide greenwash. Peabody Coal, the leading stripminer in the southwest, which had clearcut and laid bare hundreds of thousands of acres of federal and Indian land, sponsored a 10-acre tree planting in Flagstaff, Arizona. ARCO offered to sponsor Earth Day outright in Anchorage. Corporate support and influence over events of the day was so pervasive that even *Time* magazine called Earth Day 1990 "a commercial mugging." Space in the American environmental imagination had been purchased, like so many pages of advertising space in a magazine.

Denis Hayes now claims that corporate support was in fact "minimal." It will be difficult to make the same claim for Earth Day 1995, which, in the era of rampant greenwashing, promises to be even more corporate in tone and funding. Control and sponsorship of the day are essentially for sale. As this book goes to press the White House appears to be the largest bidder at $3.7 million.(See Chapter 7 for a detailed account of preparations for Earth Day 25.)

Conclusion

From the diverse ideological base I have described, American environmentalism grew to become many things—world view, life style, science; to a few, religion; and, eventually, a complex political movement with millions of followers and supporters. By the late twentieth century the environmental imagination had touched every institution in American society, and the word *environmental* had been attached to disciplines like law, biology, and ethics. Environmental philosophy and policy became the concerns of millions of Americans. In the natural and biological sciences and in tort law, environmentalism altered missions and central premises. Even the engines of the most industrialized civilization in history have been gently dusted with environmental influence.

American environmentalism emerged in the context of the most rapid economic expansion in history and matured in the technological culture that capitalism had spawned. To the extent that it has been a response to technology itself, American environmentalism has been shaped by it. And it has been shaped by capitalism as well. Although there are technophobes, neo-Luddites, and critics of capitalism in the movement, they have never been welcome in leadership circles, and their ideological influence remains minimal.

Unlike the other new social movements of the 1960s and 1970s (women's, peace, civil rights, and gay liberation), which are essentially radical, the ecology movement was saddled from the start with conservative traditions formed by a bipartisan, mostly white, middle-class, male leadership. The culture they created has persisted until very recently and hampered the success of the movement. There has always been something very safe and unthreatening about conservationists. From time to time they have aggravated ranchers, strip miners, and timber barons. But rarely have they challenged the fundamental canons of western civilization or the economic orthodoxy of welfare capitalism—the ecologically destructive system that gives the nation's resources away to any corporation with the desire and technology to develop them.

2

The Culture of Reform

We just started identifying problems and fussing around with them and coming up with a program, which typically was a regulatory program.

Russell Train, EPA Administrator, Nixon Administration

We may be "reformist" and all, but we know how to work within the context of the institutions of the society.

Michael McCloskey, Sierra Club Memorandum[1]

Despite occasional eruptions of militancy and an occasional episode of outright violence (strictly against property of course), American environmentalism has been largely a reformist movement. As the conservation crusade flourished and melded into what is today known as the environmental movement, it maintained a consistent aura of politesse. Its leaders regarded environmentalism as a sort of triangular dialogue between government, business, and national organizations. What radicalism was displayed was in the tradition of Edmund Burke—a sober examination of traditional roots rather than the spirited rebellion of a Boston Tea Party. The style was far from the confrontational irreverence of the labor, women's suffrage, civil rights, free speech, gay rights, and antiwar movements.

As the movement that took shape in the early 1960s grew and became an established voice in political debates, the prism of environmentalism narrowed to protect the economic interests of institutions that directly or indirectly supported its dominant players, the nationals. As they prospered and gained access to the sources of power, old-line conservation groups isolated themselves into a class-bound interest group, adopting a more cautious reform agenda than the movement as a whole. Leaders became caretakers of policy, and containment became a measure of success. This standard would persist as long as reform remained the primary impetus of environmentalism.

White Men's Club

As described in Chapter 1, America's conservation initiative arose in the upper reaches of society. Its founders were driven by concern rather than fear, by religious conviction rather than self-preservation. Contemporary American environmental leaders tend to be of the same bent—financially secure, well-educated, urbane professionals, too few of whom have slept in the woods, carefully read the works of Aldo Leopold, or encountered at first hand the odor of industrial pollution. In their writings and promotions they claim to be crusaders against the forces of environmental degradation, but their combativeness often consists of measured criticisms of the executive branch or periodic "report cards" on the president and Congress—with grades like C+ and C−.

It didn't have to be this way. Conservation had its fighters, and even a brief fighting era. During the mid-1940s Howard Zahniser, a former staffer at the U.S. Biological Survey, assumed leadership of the Wilderness Society from its cautious and nonconfrontational first director, Robert Sterling Yard. An aggressive lobbyist and brilliant tactician, Zahniser believed the society should form alliances with other environmental organizations to fight specific battles together. One of his first such joint projects was with the Sierra Club, which had recently come under the directorship of an impatient young mountain climber named David Brower. Together Brower and Zahniser sought to recapture the preservationist activism of John Muir and Aldo Leopold. They would do so, they decided, by preventing the damming of great rivers in the West and by drafting and lobbying for federal legislation to designate 50 million acres of land as perpetual wilderness.

Brower and Zahniser faced a showdown, not only with the government and the Chamber of Commerce, which had massive development plans for the Grand Canyon, but also with pro-reclamation advocates on their own boards. His pugnacious style and preference for activism over management eventually cost Brower his job, but not before he and Zahniser had drafted numerous versions of a federal wilderness protection bill. Scores of congressional hearings on the bills during the late 1950s and early 1960s eventually led to the Wilderness Act, which was signed by President Lyndon Johnson on September 9, 1964. Sadly, Howard Zahniser died a few weeks before the bill was signed.

The Wilderness Act mandated the federal government to preserve large areas of the country where, according to its preamble "the earth and its community of life are untrammeled by man, where man himself is a visitor who does not remain." The act created a National Wilderness Preservation System that set aside almost 10 million acres and put 5.4 million additional acres under review. The amount of land designated as wilderness under the act has since grown to over 90 million acres in almost 500 individual parcels of public land.

Brower did not go away after he was fired from his job at the Sierra Club. But he did leave one of the most aggressive organizations in the conservation movement in the hands of more conciliatory leaders. Still honored on birthdays and on the anniversaries of undeniable triumphs, Brower (dubbed the "arch-druid") served on the Sierra Club Board for years, but was kept at arm's length by the new guard.

By the late 1960s, mainstream environmental activism had become a respectable profession, the environment itself an intellectual battle ground. The professional environmentalist's major activity was wrestling with corporate and government officials over legal subtleties and regulatory standards. Mainstream environmental boards, like the boards of most establishment nonprofits, have changed little since those days. They remain mostly male, white, and patrician. Board members of either gender are generally selected for their connections to money or power, or both. When minorities, non-wealthy women, Native Americans, and activists expressed displeasure at their exclusion, they were given token representation on some boards.

Mainstream power resides to this day in a "white men's club," a reality reflected in the agendas and priorities of the national environmental organizations. In 1992 the Public Liaison Office of the EPA surveyed the 15 largest environmental organizations in the country to determine their priorities and how they are selected. By counting the occurrence of themes in organizational literature and conducting focus interviews with organizational leaders, EPA learned that the mainstream groups' highest priorities were ecosystems, environmental education, and environmental laws and legislation. Their lowest priorities were toxic wastes, human health, and technology.[2] To no one's surprise, the priorities of the organizations were set by the boards of directors.

Achievements

When you've been climbing a mountain for a long time and the summit seems to be as far away as ever, it can give you new heart for the ascent to look back and see how far you've come.

Harold Gilliam, Earth Day 1990

The early achievements of the nationals were significant. By educating vast memberships and the public at large, mainstream organizations placed the state of the natural environment near the top of the American agenda and kept it there for 25 years. Through the 1960s and 1970s their organizing and lobbying efforts compelled even the most reluctant politicians to take action on behalf of the environment. By signing the Wilderness Act in 1964, Lyndon Johnson established a global precedent for industrial nations to set aside large roadless areas simply to keep them wild. In doing so, he aggravated some of his own powerful supporters in the mining, timber, oil, gas, and cattle industries. Richard Nixon, no friend of the harp seal or the redwoods, nonetheless signed the National Environmental Policy Act (NEPA) into law on January 1, 1970. He was responding to polls that showed strong public support for federal protection of environmental health. In his State of the Union Address, delivered three weeks later, Nixon described environmentalism as a new form of selflessness. He called for the nation to "make peace with nature" and repair "the damage we have done to our air, to our land and to our water." *Time* magazine called the environment "Nixon's New Issue."

NEPA, which institutionalized the requirement for environmental impact statements on new building projects, mandated all federal agencies to protect the environment. The Environmental Protection Agency (EPA) was created and the Council on Environmental Quality (CEQ) met for the first time on Nixon's watch. When he signed NEPA, the president declared that "the 1970s must absolutely be the years when America pays its debt to the past by reclaiming the purity of its air, its waters, and our living environment. It is literally now or never."[3]

Nixon never claimed to be the "environmental president," and he backtracked on many of his own environmental initiatives. Yet he launched "the environmental decade"—with an assist from William Ruckelshaus and a major push from the emerging environmental lobby that represented a vast bipartisan voting constituency.

Over the next 10 years, 23 federal environmental acts were signed into law. The principal ones are the Clean Air Act of 1970, the Occupational Health and Safety Act (1970), the Water Pollution Control Act (1972), the Marine Protection Act (1972), the Federal Insecticide, Fungicide, and Rodenticide Act (1972), the Marine Mammal Protection Act (1972), the Endangered Species Act (1973), the Safe Drinking Water Act (1974), the Toxic Substances Control Act (1976), the Resource Conservation and Recovery Act (1976), the Clean Water Act (1977), and the Comprehensive Environmental Response Compensation and Liability Act ("Superfund," 1980).

As a result of these acts and other initiatives of the environmental lobby, tens of millions of acres have been added to the federal wilderness system, environmental impact assessments are now required for all major developments, some lakes that were declared dead are living again, and clean air and water have become virtual human rights. While American air and ground water are still nothing to brag about, they are undeniably less polluted than they would be without the goading of organized environmentalists. The latter can also take credit for the lowered lead content of the American bloodstream, a 75-percent reduction of PCBs in American fatty tissue, and the virtual elimination of Strontium 90 from cow's milk and children's bones. In addition, several species of animals have been saved from extinction.

As a cultural and political phenomenon, the environmental movement of the 1970s changed the way millions of Americans perceived their relationship to the Earth. Words like *ecosystem, biosphere,* and *planetary citizenship* have become familiar household concepts and recycling a civic duty. "Ecology," one politician quipped, "became the political substitute for motherhood." The word *environmental* was attached to disciplines like biochemistry, architecture, and engineering; new departments at major universities hatched a host of advocacy scientists to join the national organizations and the new federal bureaus.

During the 1970s the environmental movement became a political force to be reckoned with. Attempts were made to recruit millions of card-carrying "enviros" into a powerful national voting block—a broad new bipartisan constituency spanning every imaginable ideology and party. Pollsters began to advise candidates that the "environmental vote" would be a major factor in elections. Candidates who had never recycled a beer can, slept in a tent, or seen a deer outside of the zoo declared themselves environmentalists. A significant number were elected to both sides of the aisle, and a bipartisan environmental legislative strategy became feasible. In the years before the election of Ronald Reagan new environmental issues were codified into laws and regulations and a new species of lawyer evolved to enforce them.

Sue the Bastards!

Sue the bastards!

War cry of the Brookhaven (N.Y.) Town Natural Resources Council

During the osprey hatching season of 1967 a group of bird watchers in Stony Brook on the north shore of Long Island met to brainstorm the best way to protect the local osprey from insecticide poisoning. Rachel Carson had convinced them in *Silent Spring* that DDT, sprayed to eradicate mosquitoes, was finding its way into the osprey's food chain and rendering eggshells so brittle embryos could not survive long enough to hatch. They formed an environmental discussion group called the Brookhaven Town Natural Resources Coalition. The BTNRC, which included scientists and lawyers, rented an office over the Stony Brook fire department. In early discussions about what must be done to stop the spraying, someone said "sue the bastards," and attorney Victor Yannacone rose to the occasion. The coalition sued the Suffolk County Mosquito Control Commission and won an injunction against DDT spraying on local marshes and ponds. The suit spawned a nationwide attack on DDT, which culminated five years later in an outright ban by the EPA.

Yannacone approached the National Audubon Society with a proposal to open a legal department. It was a strange choice of organizations. The ultracautious and conservative society had played an exceedingly passive role during the Rachel Carson controversy. Audubon directors with close ties to chemical companies, particularly chairman Gene Selzer and general counsel Donald Hays, rejected the proposal. Yannacone decided to go it alone. In 1967 he established the Environmental Defense Fund (EDF) with a startup grant from the Ford Foundation.

Yannacone's dynamism and vehement oratory convinced other environmentalists that litigation could become a powerful tool in their struggle. Before long two more environmental law firms, the Natural Resources Defense Council (NRDC) and the Sierra Club Legal Defense Fund (SCLDF) were opened, with generous assistance from the Ford Foundation. Law schools across the country began offering courses and, eventually, full curricula in environmental law.

The NRDC was begun by a handful of Yale Law School students, led by James Gustave "Gus" Speth, who sought more challenging and socially redeeming careers than corporate law firms could offer. Earlier they had formed a group called the Legal Environmental Assistance Fund (LEAF) and applied to the Ford Foundation for support of a permanent project. The foundation, which had recently been investigated by a congressional committee for granting money to EDF and the Oceanville-Brownsville school experiment, was reluctant to fund another bunch of unknown and potentially radical young lawyers.

Foundation officials did, however, introduce Speth and his colleagues to Republican Wall Street lawyers Whitney North Seymour, Jr. and Stephen Duggan. Along with David Sive ("the father of environmental law"), Seymour and Duggan had just won the landmark Storm King case stopping Consolidated Edison from building a massive pump storage plant on the Hudson River. In a marriage arranged by the foundation, Laurance Rockefeller joined Seymour and Duggan on the board of NRDC, which was staffed by the young Turks from Yale. When the Nixon administration, through the IRS, tried to obstruct NRDC's tax exempt status, pressure from Republican supporters of the project, including several former presidents of the American Bar Association, smoothed the way.

For the second time, Ford's matchmaking had joined the energy and ardor of passionate young lawyers with the cautious pragmatism of respectable legal veterans.[4] By 1970 New York had two environmental law firms, EDF and NRDC. A third group, the SCLDF, also started with Ford Foundation seed money, was headquartered in San Francisco. Established as the legal arm of the Sierra Club, SCLDF later broke away from the club's conservative and obstructionist leadership to become an independent firm.[5]

EDF and NRDC differentiated themselves from SCLDF by hiring scientists—mostly biologists, chemists, and economists—to assure themselves of in-house experts who could help prepare law suits. SCLDF, realizing that they would have to hire and pay independent scientific expert witnesses even if they employed their own, chose to form a lawyers-only organization. Unlike NRDC and EDF, SCLDF represents litigants but does not initiate lawsuits on its own. Thus a small grassroots organization bent upon suing a "bastard" over a timber sale or a toxic dump could turn to SCLDF.

Branches of the new law firms were opened in Washington, Denver, Oakland, Seattle, San Francisco, and Los Angeles, and "sue the bastards" became the war cry of the environmental bar. Government scofflaws and corporate polluters across the country found themselves in court or hailed before administrative review boards. Before the end of the 1970s the state of California had been forced to reduce lead emissions in gasoline, the manufacturers of TRIS—a cancer-causing fire retardant used in children's' pajamas—had been enjoined to cease production, and successful suits had been filed against the supersonic transport (SST), pesticide abuse, and lead toxicity. Over 40 suits were filed to enforce the Clean Air Act (35 of them by the NRDC).

Aggressive litigation worked for quite a while. One of its most noteworthy achievements was establishing the principle that nature, flora, and fauna had rights ("standing") and were entitled to the same legal protections as corporations and oppressed human beings. Arguments on this issue delayed construction of the Alaska pipeline for more than three years and completely stopped construction of a barge canal across Florida. While environmental advocates brought suits to federal courts to enforce federal legislation, a host of private attorneys worked at the local level, often *pro bono,* to fight environmentally unsound land development. "Environmentalism," David Sive has declared, "used litigation as no other social movement has before or since."[6]

In the beginning of the 1970s, environmental litigators believed that most of their suits would be filed against corporate polluters and the extractive industries and that the federal government would be their friend. They assumed that the EPA would issue pollution regulations giving priority to public health and safety; that the Fish and Wildlife Service would enforce the Endangered Species Act; and that the Forest Service would protect watersheds and manage the national forests for recreation and wildlife. They expected the Justice Department to prosecute criminal and civil violators of environmental laws vigorously and the federal courts (including the Supreme Court) to scrutinize government actions and interpret federal laws impartially. How wrong they were, on all accounts!

Since then, environmentalists have joined a long list of American political activists who have realized from bitter experience that the government does not enforce its own laws, and that citizens and citizen organizations must do so themselves through litigation. (If Gifford Pinchot had foreseen this situation, the federal conservation apparatus might be very differently designed.) In the 1970s the dockets of the courts were stuffed with cases brought by the EDF, NRDC, SCLDF, and the National Wildlife Federation against the federal government, which turned out to be the country's leading environmental scofflaw.

The litigative strategy paid off as long as federal judges continued to rule in favor of the environment, which occurred with some regularity until Ronald Reagan began making appointments to the bench. For the most part, the founders and financial supporters of the legal environmental movement were pleased with the activities carried out by the new breed of lawyers they had created. There was, however, the problem of certain overaggressive young lawyers who focused their zeal on corporate environmental abuses. It was litigation against these corporate violators that landed Victor Yannacone and his colleagues in deep trouble with their own trustees and funders, especially the Ford Foundation. They were simply too adversarial in their rhetoric, too confrontational in court, and too successful against companies with close associations to foundation trustees. The "sue the bastards" style had to go, and so did Yannacone. He was fired from EDF in 1970.

Thereafter, Ford remained committed to supporting environmental litigation but imposed certain restrictions. For several years all cases being considered by environmental law firms funded by Ford had to be screened by the "five gurus," an internal committee composed of former presidents of the American Bar Association.[7] In addition, EDF was required to form, with Ford's review and approval, its own bipartisan litigation review committee, providing foundation executives an additional layer of protection from embarrassing associations with confrontational litigants.

Not surprisingly, environmental litigation gradually became less and less aggressive, settlement and deal making became popular, and the fighting spirit that characterized the early days disappeared. NRDC veteran Al Meyerhoff, however, still believes that there is hope for litigation: "I'm still suing and I'm still winning," he said in September of 1994, "and we're getting some better judges on the bench."[8]

Despite Meyerhoff's optimism, environmental law today has a completely new look. Law schools offering environmental curricula now attract students who expect to defend corporations against NRDC rather than work for NRDC. Every year since 1970 the Environmental Law Center in Washington has sponsored a convention of American environmental lawyers. During the seventies and early eighties flannel shirts and hiking boots were the usual attire. Last year, remarked one veteran SCLDF lawyer, "95 percent of the attendees were corporate compliance lawyers. Off in a corner was a small circle of guys without ties drinking Bud out of the bottle. I never felt more lonely at a convention in my life." "And these guys are ready for war," he added. "Private corporations will put 20 lawyers to work on a case and break your balls. Take Noranda Mines up in Cooke City [Montana] for example. They've hired most of the legal talent in Montana . . . and they're all calling themselves 'environmental lawyers.' How can we possibly stop that mine?"[9]

Science

At base, the movement has scientific integrity or it has nothing.
Jon Roush, Executive Director, Wilderness Society

Environmental pollution is an almost incurable disease, but it can be prevented.
Barry Commoner, Director, Center for the Biology of Natural Systems

Environmentalism depends heavily on science. Participants on both sides in environmental debates resort to scientists as arbiters and experts. Ultimately the great controversies over toxic pollution, global warming, risk assessment, and biodiversity will be settled by science and scientists. But not without a lot of posturing and professional insults.

It is difficult for laymen and journalists to assess the veracity of environmental scientists. On the one hand, they are presented as objective observers, schooled in the scientific method and expected to seek the truth in nature. On the other hand, most of them are paid to cull whatever evidence employers or contractors need to make their points and win their arguments about the state of the environment. This is as true for scientists employed by Monsanto as for scientists working at EDF.

Rachel Carson forewarned that industry will always be prepared to keep regulation at a minimum by purchasing more scientific expertise than the movement. Moreover, paid experts will always be able to persuade enough elected representatives that environmental threats are being exaggerated by "self-interested" activists. Thus the effectiveness of using the findings of environmental science as the principal tool of reform environmentalism must be constantly challenged.

At the same time, we need to keep in mind a few undisputed findings of environmental science as we debate the subject. The most important are these: lead particulates have been impressively reduced in the atmosphere; DDT is no longer found in American body fat, which also contains considerably fewer polychlorinated biphenyls (PCBs) than it once did. Mercury has virtually disappeared from Great Lakes sediment; and Strontium 90 is no longer found in either cows' milk or mothers' milk. What all these facts have in common is that they are the result of outright bans against the use or production of the substances in question.

Other toxic chemicals still found in the natural and anatomical environments are those that government is attempting to control or limit through regulatory standards. For most of these substances we are at best holding steady. "When a pollutant is attacked at the point of origin in the productive enterprise, it can be eliminated. Once it is produced it is essentially too late," according to Barry Commoner, founder and director of the Center for the Biology of Natural Systems at Queens College, City University of New York.[10]

Although this conclusion has yet to be accepted by the American government, industry, or even by many in the American environmental movement, it is the constant rallying cry of independent environmental scientists. Commoner is widely respected by environmental scientists around the world, but his thesis is largely ignored by American politicians and environmentalists. On the subject of pollution, he is essentially a socialist, and socialism simply doesn't sell in the American political domain where mainstream American environmentalists have chosen to focus their struggle against pollution. There they are faced with what Commoner calls "a taboo against social intervention in the production system."[11]

Money Talking

You can't run strategy to save the planet with the mind of a fundraiser.
David Foreman, Executive Editor, Wild Earth Journal

The best work I've done was unfunded. No one would touch it. That's how I knew it had to be done.
Robert Bullard, Clark Atlanta University

In 1993, about 2.5 percent of the $126 billion contributed to various causes went to what can be broadly described as environmentalist organizations. The $3.12 billion donated in that category in 1992 also includes contributions to animal welfare, nature conservancy, and other apolitical activities. The active environmental movement described in this book took in an estimated $2.5 billion, more than double what it received in 1987, the first year the American Association of Fundraising Council kept account of donations. In 1965 the ten largest environmental organizations in the country ran on less than $10 million; by 1985 that number reached $218 million; by 1990, $514. Today total funding is approaching three quarters of a billion dollars. Almost 70 percent of the total is absorbed by the twenty-four organizations that comprise the Washington-based mainstream sector of the movement.[12]

In a world where "one member, one vote" is replaced by "one dollar, one vote," money talks.[13] In the mainstream environmental movement, money also sets agenda, strategy, targets, and priorities. Whether it is direct mail money, large individual donations, or foundation and corporate contributions, the money tends to say "reform." Moreover, big money talks in a voice very different from that of pocket change. Foundations and six-figure donors, who act as sort of progressive venture capitalists by donating almost $200 million a year to environmental causes, play a vital role in shaping the environmental agenda. This becomes more true as the nationals are forced to rely more heavily on large donors by declining responses to direct mail requests. (See Chapter 7 for a fuller discussion of declining memberships.)

Mainstream organizations have been drawn into this dependency by their need to survive and the dictates of philanthropy. All four major sources of environmental philanthropy in the United States—direct mail, large donors (including bequests), foundations, and corporations—set the parameters within which organizations must function to survive. Thus American environmentalism is both defined and limited by the philanthropy that supports it. Although that isn't necessarily a bad thing, it can be very restrictive.

Of course, this situation prevails elsewhere as well. In the Netherlands, and to a lesser extent in other European countries, environmental organizations receive government support, which creates its own kind of problems and limitations. In the U.S. county governments occasionally fund local citizen organizations, but there is very little public financial support for environmental activism, particularly when government itself is the target.

Mainstream environmental leaders generally deny that money influences their decisions or priorities, or that they are in any way beholden to their funding sources. It's a claim similar to the media's denial of influence from Madison Avenue. It isn't true, but it has to be said to preserve an aura of independence and objectivity. The fact is that almost all funding comes to the nationals with an implicit, if not explicit *quid pro quo*. Perhaps the only money without strings is the small amount earned by the Sierra Club, Audubon, Wilderness Society, and National Wildlife Federation from retailing of tee shirts, posters, mugs, calendars, and coffee table books.

This situation induces organizations soliciting the public for support to select issues appealing to donors, to overstate the problems, and to exaggerate their triumphs in attacking them. They are likely to favor more fundable "designer issues" that are popular at the moment (baby harp seals, Alaska, environmental justice) but that often don't go to the heart of the organization's mission or even to the requisites for a healthy environment.

Direct Mail

The United States is one of the few countries in the world that subsidizes bulk mail marketing and provides even lower rates for nonprofit organizations. Second and third class nonprofit postal rates are so low that direct mail has become the preferred method of fundraising and membership building. There is simply no faster or more efficient way to recruit a nationwide membership in the thousands or, in the case of many environmental nationals, the millions.

Direct marketing methods have paid off in spades for the mainstream environmental movement. Mail campaigns are particularly well suited to environmental issues, with their broad demographic allure and emotional appeal. By focusing attention on the ample supple of major and minor environmental catastrophes, direct mail allows the movement to offer millions of people who deeply fear a loss of health and freedom from environmental insults a sense of empowerment.

The environmental movement came late to direct mail. Until the early 1980s it relied on foundation grants, canvassing, and member-to-member solicitation. Low-cost direct marketing has permitted the movement to expand its base of financial support exponentially, to create a new political constituency, and to educate millions of people about environmental issues. Ultimately, according to journalist, political consultant, and direct mail copy writer Jeffrey Gillenkirk, "direct mail became the movement's heroin."[14] And, like any addictive substance, it began to control the life of its user.

In an electoral democracy, where a few million sensitized voters can swing a local or national election, membership organizations demand and receive political attention for their cause. Membership building has thus become the American way of political organizing. The problem is that when direct mail becomes the preferred method of building membership, large national organizations all too often come to regard it as the only form of political organizing needed. The results of direct mail campaigns then become tacit opinion polls that organizations rely on to guide policy and programs.

Focus group surveys of "attached and casual contributors" to environmental causes conducted for the League of Conservation Voters (LCV) by Environmental Opinion Studies (EOS) show direct mail donors to be a cautious and conservative group. They "approve of lobbying activities, but believe that the most effective position for an environmental organization, vis-à-vis the federal government should be 'neither adversarial nor sycophantic.' " EOS concluded that their findings contradict the "impression that environmentalists are too extreme, too confrontational or too narrowly focused."[15] In other words, moderate reform works in the mail, and organizations that project a confrontational or extreme image will have to seek other means of winning support.

Direct marketing is an unfortunate expedient for environmental organizations, whose leaders rarely discuss it in public. Their embarrassment comes not only from the public's general repugnance of "junk mail," but also from the fact that direct mail is an expensive, wasteful, and environmentally degrading way to raise money. Anyone who has seen a bulk mailing being loaded onto trucks with fork lifts has a sense of how many trees had to be ground into pulp to get that letter, brochure, order card, and return envelope to two or three million people. Although some recent mailings have been printed on recycled paper, most of the stock used has a low post-consumer content. Slick four-color brochures printed on heavy coated paper rarely if ever contain more than 20 percent recycled fibers (nor do the monthly publications of the Audubon Society, Sierra Club, Wilderness Society, or National Wildlife Federation[16]). And anyone who has studied direct marketing techniques knows that millions of envelopes are discarded unopened. Nonetheless, for environmental organizations striving to survive in a competitive world, money means members and members mean direct mail. An estimated 65 to 70 percent of the combined budget of Washington-based mainstream environmental organizations comes through the mail from millions of small donors.[17]

Direct mail consultants (who themselves thrive on volume) advise their clients that the only way to maintain a stable or growing membership through direct solicitation is to "stay in the mail." There must always be something on the way to members or prospective members, whether a membership solicitation, an invoice, a gift solicitation, a premium, a renewal notice, or an advance renewal notice. Ironically, the characteristics that earned direct mail the nickname "junk mail" are often those that "pull" the best results—a heavily printed No.10 outer envelope, a six-page, single-spaced letter, an order card with a punch-out token, a business return envelope, and a glossy four-color brochure.

Membership organizations generally experience an attrition rate of between 30 and 40 percent a year. Whatever the reasons, many members just do not renew. Even to maintain a constant membership, environmental organizations have to mail hundreds of thousands, if not millions of prospect letters—"cold solicitations"—every year. Moreover, direct marketing is a constant, ongoing process that requires a professional well-paid staff to determine what will and will not work. They must be backed up by pollsters, list managers, consultants, and at least one contracted fulfillment service.

Direct mail managers and outside consultants are often present when environmental organizations are discussing long-term strategy. They frequently influence it simply by reminding their clients of what has or has not worked in the mail. No matter what else a membership organization dependent on direct mail may do right, if it disobeys the rules of direct marketing or hires incompetent mail managers, it will eventually fail; direct mail has become the lifeblood of the mainstream movement.

Direct mail is not solely a fundraising, membership-building tool; whether intended to be or not, it is an educational medium. The focus survey performed for the League of Conservation Voters indicated that recipients of direct mailings "count themselves among those who need to be educated" and rely on the information received in the mail to learn about "issues, activities and opportunities." Contributors told surveyors that they often "forgive long fundraising letters because they contain . . . real and substantial information." As one San Francisco donor said, "I'm going to give money in order to get information."[18]

While a campaign of a million pieces may recruit only a few thousand new members, another two or three hundred thousand people will open the outer envelope and read the contents. From these promotional materials, often inspired by pollsters and always written by professional copy writers, recipients learn about the organization's priorities and, by implication, those of the movement as a whole. If six mainstream organizations decide simultaneously that talking about the drilling of oil in the Arctic National Wildlife Refuge (ANWR) is the best way to raise money (as happened in 1991), then millions of potential contributors will get the impression that ANWR is the most important national environmental issue.

Well is it? And if it is a high priority, is it because the breeding grounds of some furry mammals are threatened or because it reinforces the industrial world's insatiable appetite for oil? Is the Amazon rainforest a significant global environmental issue because a lot of beautiful birds exist there or because the ancient rainforests are the lungs of the planet? The direct mailings about the ANWR focused on the effect on caribou and, except for the Greenpeace and Friends of the Earth's campaigns, barely mentioned oil dependency. Even though rainforest mailings mention the carbon-absorbing properties of ancient forests, their emotional appeal is to love of birds and wilderness aesthetics.

Some direct mail consultants, most notably Craver, Matthews and Smith in Washington, have contracts with several environmental organizations. Roger Craver and other direct mail specialists are often called into top management meetings at Washington environmental organizations, particularly in crisis situations. And their cautionary advice about what will and will not affect the outcome of ongoing direct mail solicitations is heeded. The essence of their counsel is that to succeed in the political direct mail business you have to win, look like you're winning, or look like you're going to win. "You don't really have to win to win," is how Patti Schifferle, a former employee of the Wilderness Society, describes it. "You can look like you're winning if you compromise, but not if you are losing."[19]

Even Greenpeace, for years the purest of the nationals, which recruited its first million or more members through a door-to-door canvas, is not immune to the seductions of direct mail. During the late 1970s and early 1980s its "Kiss This Baby Good-bye" campaign, featuring a close-up of a baby harp seal with eyes like Kim Basinger's on the outer envelope and a commitment inside to confront the seal hunters, pulled in thousands of members and millions of dollars. The organization was, naturally, committed to saving the harp seal, a worthy humanitarian cause, though not one of great ecological significance. That campaign also committed Greenpeace to a policy of direct confrontation, a tactic that worked in the field for over a decade. It apparently also worked in the mail, as the organization's membership outpaced that of all other environmental organizations.

Then, for reasons that are still being widely debated in the movement, Greenpeace began losing membership.[20] Between 1990 and 1994 Greenpeace USA lost almost a million members and suffered a 40 percent reduction in revenue, from approximately $5 million to $3 million. When response falls off in direct mail, it is difficult to determine whether it is caused by "list fatigue"—overuse of the same mailing lists—or by a subtle preferential shift in the market for social and political causes—a shift that might be described as "donor fatigue."

The response of direct marketers to decreased front-end response is to test. Greenpeace decided to test the central premise of the organization with a direct mail campaign mailed out in an envelope labeled "Confrontation Works." It was a bit of a risk, given the fact that the LCV focus group surveys had indicated that direct mail donors were put off by strident or confrontational rhetoric. Still, Greenpeace has always appealed to a slightly more "radical" donor base than mainstream organizations, and the letter inside was a tribute to the many nonviolent direct actions that had been so effective against whalers, seal hunters, and nuclear testing. The test "bombed." That is, it pulled such a low percentage of returns that it could not make money if mailed to the entire universe of potential Greenpeace members. It was never mailed.

Greenpeace, as other environmental organizations would have done, read the result as not only a negative market test but also a poll. Of course, it was a poll not of members but of potential members. Never mind. What it said to Greenpeace was that of the several million people out there who might become supporting members of the organization, very few were impressed with direct confrontation as a tactic—even when it worked in the struggle against environmental adversaries. Significantly, Greenpeace subsequently became a much more low-key organization, and is often derided by former supporters for having gone "corporate," "soft," and, perhaps worst of all, becoming "mainstream." This is not an indictment of Greenpeace, which remains in many respects the most effective environmental organization in the world. It is, however, an indictment of direct mail dependency.

Sometimes an organization will not even bother to test a concept or potential campaign, but will be talked out of it by direct mail consultants. This happened at Greenpeace in 1993 when its leaders wanted to fund a campaign in support of community opposition to a massive waste incinerator in East Liverpool, Ohio (see Chapter 7). For Greenpeace the campaign would have been an important step into the movement for environmental justice. Roger Craver persuaded the leaders not to take it, arguing that the subject simply wouldn't work in the mail.

Direct mail consultants like Craver, Matthews and Smith make money when their clients make money. If they believe a campaign will not pay off they oppose it, no matter how critical the problem. And if Greenpeace or any other organization can't raise money to conduct an action or campaign, it may not be conducted. So if 2 or 3 percent of an organization's marketing universe (an above-average response) joins in response to direct mail package A, while package B bombs (pulls less than half of 1 percent), the organization is motivated to follow the tactics in A rather than B. Both A and B may be worthy causes, but if B is of greater ecological consequence, it won't matter. Tactic A will prevail.

Another problem with the expedient priorities reflected in direct mail promotions is that they subtly mislead the entire environmental constituency. Asbestos, acid rain, and pesticide residues are all real and potentially disastrous problems. But they have all been, at one time or another exaggerated, overexploited, and given misplaced priority by mainstream organizations groping for public outrage, a quick PR triumph, or productive envelope copy. When a direct mail campaign works, it is repeated over and over until it stops working, and the subject around which it is created eventually becomes an organizational priority—regardless of its intrinsic merit or environmental significance.

Once a soliciting organization has used an environmental problem to raise a few million dollars, it is obligated to start a program or open an office to solve the problem. Because birds and mammals elicit more sympathy than lower life forms and therefore work well in the mail, they become high priority critters, although it can be argued ecologically that less charismatic endangered plankton, frogs, conifers, and dung beetles are more critical to survival of the biosphere.

To sustain a large membership through direct mail an organization must project success. "Results," declares Fred Krupp of the Environmental Defense Fund, "is all my members want out of us." Thus a careful reading of direct mail copy from mainstream organizations reveals that many groups take credit for the same triumph. This is particularly true among the legal groups, where one organization, say the Sierra Club Legal Defense Fund (SCLDF), litigates a case against a government agency or a corporation, while half a dozen other organizations support the suit with research, an *amicus curiae* brief, or a little money. If you judged by the next newsletter or direct mail solicitation of any of the organizations you'd be convinced it won the case single handed.

Take, for example, the case of the marbled murrelet, a species of bird endangered by clearcut forestry in the northwest. At the request of 12 separate clients SCLDF brought a suit in defense of the murrelet and won. "The most each of those clients kicked in," according to SCLDF lead attorney Vic Sher, "was a few hundred dollars to help defray out-of-pocket expenses. We put in tens of thousands of dollars worth of attorney's time, expert witnesses, etc. Yet all 12 plaintiffs declared victory in their promotional materials, none of them mentioning that SCLDF was lead attorney. Sher finds that his smaller clients are the most appreciative and willing to attribute credit where it is due. "It's the large organizations that seem bent on taking all the credit," he says, diplomatically avoiding specific names.[21]

For better or worse direct mail is wearing out for the environmental movement. "People have been over-dunned," quips copywriter Jeff Gillenkirk. What's next? "800-numbers placed in nature documentaries and interactive cable are possibilities," says Gillenkirk.[22] Meanwhile, the nationals have resorted to the sorriest extension of direct mail fundraising—the New York boiler room of $6.00-an-hour telemarketers hustling advance renewals or emergency funding—teenagers reading almost phonetically from index cards or a "talk book" about some spurious crisis requiring "your immediate attention."

Foundations

If a foundation has a large interest in Alaska and a lot of money, you definitely have a large interest in Alaska.
Bill Turnage, Former Director, The Wilderness Society

Foundations are laced with people who don't want to awaken the people.
Stewart Brandborg, Former Director, The Wilderness Society

Overall, foundations provide about 7 percent of the total mainstream environmental budget. Some organizations rely more heavily on them than others; these groups tend to be more conservative, more inclined to take the "soft path" of environmental reform. Mainstream environmental organizations are safe places for foundation philanthropy because they obey the unspoken dictum: "Do nothing to jeopardize the value of your benefactor's endowment." Most of the nationals have been handsomely rewarded for their obedience from the fortunes of the Ford, Rockefeller, Pew, Stern, Mott, W. Alton Jones, and Kendall families, whose financial overseers are particularly sensitive to the economic orthodoxies of corporations. The message is clear, though rarely uttered: be cautious reformers, challenge specific violators, take them to court. Lobby for environmental regulations. Educate the public. But don't rock (or knock) the capitalist boat if you intend to rely on significant foundation funding.

Because foundation grant money tends to be project specific it has a lot more policy leverage than membership money, which is allocated to general support. Although membership solicitations often mention general strategies and specific projects the organization intends to pursue, they are not contractually obligated to carry them out. Not so with foundation money. Foundation officers and staff impose constant oversight, requiring frequent reports, on-site visits, and detailed audits.

Recently some large foundations have been even more proactive in their philanthropy. Believing that the environmental movement was drifting, they have begun designing their own strategies and tactics and offering organizations money to carry them out. Foundations such as W. Alton Jones, the Pew Charitable Trust, the Rockefeller Family Fund, and the Nathan Cummings Foundation have been meeting to decide where the environment movement should be going. They create multimillion-dollar mega-projects and invite organizations to apply for grants to activate them.

Mainstream organizations desperate for resources are rising to the occasion, without concern that money is shaping their policies and strategy. When the W. Alton Jones Foundation declared that "grassroots organizing" was the future, mainstream organizations that hadn't spent a minute organizing locally stepped up to the window with proposals that read as if they were prepared for the task.

One particularly controversial proactive campaign was designed by the Pew Charitable Trust, the largest foundation funder of environmental projects in world. In 1993 Josh Reichert, environmental director of Pew, circulated a proposal to other foundations to join Pew in funding a new environmental institution he was calling Environmental Strategies. Its mission would be "to assist environmental organizations to conduct public education campaigns on priority national environmental issues." Implicit in its design is a critique of the environmental establishment's communications and lobbying abilities and an assumption that the Green Group has failed to fulfill its essential purpose of "identifying public policy issues that are of greatest mutual concern and importance."[23] The most controversial aspect of the project is that the board would be composed of both environmentalists and grantmakers.

Reichert's aim was to create a new, collaborative organization based in Washington (of course) "to improve the campaigning skills of national, regional and local environmental groups and to help see to it that those skills are employed to the best effect on priority national issues." Although Reichert never states explicitly who will determine the priority issues, he points out that "foundations play an integral role in the planning and execution of issues campaigns because, as often the sole source of funds for broad public outreach, grantmaking becomes the focal point for inter organizational cooperation. Consequently, unless foundations take the initiative in collaborating with national and local environmental organizations, little improvement can be made in campaigning on national environmental issues." Reichert adds with emphasis that Pew would "*never* ask an environmental organization to change its agenda," though he acknowledges that they very well might do so to get the money dangled before them.[24]

The irony of Reichert's proposal is that it calls for using strategies borrowed from antienvironmental organizations, which initially coopted *their* most effective tactics from the environmental movement. "Special interest politics has spawned the career political consultant and the orchestrated issues campaign." says Reichert.

> For considerable sums of money, public opinion can be molded, constituents mobilized, issues researched, and public officials buttonholed, all in a symphonic arrangement. There are media spots, direct mail drops, phone banks, and old fashioned lobbying, tactics employed in specific target areas, all informed by opinion research. While business and industry has made extensive use of them, environmentalists have been slow to employ and, equally important, to coordinate these new political arts. As a result environmentalism has fallen behind in a political arms racc that requires even higher levels of organized constituent involvement to influence officials and engender administrative or legislative action.[25]

Reichert sees Environmental Strategies as a service bureau for both grassroots and national environmental organizations. He hopes that its existence will "institutionalize a relationship with grassroots organizations seeking to project a presence in Washington."

The main problem facing Reichert's scheme is lack of leadership. When I spoke with him in the summer of 1994 he had interviewed and considered scores of candidates for the top job at Environmental Strategies but found none of them suitable. I suggested a few candidates. He had either already rejected them or thought them inadequate to the task. "I don't want someone who knows the facts, or can articulate them persuasively, I want someone who wants to win and knows how," said Reichert. Who would be his dream applicant for the job? "James Carville," he said without a second's hesitation.[26]

Many organizations are insulted by Reichert's implicit assertion that they are failing or incompetent, particularly when he says things like "we started this because no organization was doing what we thought needed to be done." Reichert's creation of Earth Force, a $12-million project that provides ecological education to youngsters with the hope of creating environmental activists for the future, ruffled some feathers. Several mainstream organizations, notably the National Wildlife Federation, had similar programs under way, as did non-environmental groups like the Boy Scouts. To Reichert, who hired a temporary staff of researchers and consultants to assess the field, none of these programs quite measured up.

While they bemoan the intrusion of outsiders in their business, mainstream organizations line up for the money being offered by Pew and other proactive foundations, which report that grant proposals flood their offices any time they so much as hint at a new idea. When Reichert began talking about a national forest campaign to save the remaining 5 percent of old-growth forests in the U.S. his office was inundated with proposals from mainstream and grassroots organizations alike. According to Reichert, most of them proposed doing exactly what he said needed to be done.

The very existence of a few million dollars earmarked for forestry reform has caused a donnybrook inside the movement between reform organizations like the National Wildlife Federation, the Sierra Club, and the Audubon Society, on one side, and more activist grassroots organizations like the Native Forest Council and Heartwood on the other. Grassroots organizations, in particular, fear that proactive funding will load the forest management debate in favor of incrementalists like NRDC, the National Wildlife Federation, and the Sierra Club, who endorse ecosystem management schemes that permit controlled logging in national forests. "Foundation money behind a compromise position tempts nonprofits to moderate their hardline stance or risk being left out of the coalition," according to Tim Hermack, Director of the Native Forest Council, which advocates a zero-cut policy in all public forests.[27] In fact, that is exactly what has happened to the Native Forest Council. It, and other grassroots organizations that believe no more of the remaining old-growth timber should be cut, have been kept out of all Pew-funded deliberations on ancient forests.

Perhaps surprisingly, Pew President Rebecca Rimel doesn't entirely disagree with Hermack. "One of the dangers of this kind of funding," she says, "is that foundations have this wonderful way of not only knowing what the world's problems are but how to solve them." Pew's board members "were very concerned that buried in this proactivity or strategic thinking would be an arrogance, a sense that we didn't know how to listen, we weren't going to be responsive to grantees, not only to their needs but also to their expertise and that it was going to be our way or no way." [28]

Foundations, like the general public, are fickle and just as subject to the herd mentality. When a handful of foundation directors decide that one environmental strategy is more worthy than another their colleagues tend to follow. If six foundations decide to pull out of the mainstream environmental movement altogether and shift their philanthropy to the grassroot levels—funding communications infrastructure, organizational development, special training, or whatever they perceive to be vital—the character and energy of American environmentalism will change overnight.

Corporate Philanthropy

People charge Mr. Rockefeller with stealing the money he gave to the church . . . but he has laid it on the altar and thus sanctified it.

Anonymous Parson, ca. 1910

While wanting to serve the company's ''social responsibility" each of [our] grant-making programs really serves our self-interest as well.

Roy Marden, Manager of Corporate Relations, Phillip Morris [29]

"The big corporations, our clients, are scared shitless of the environment movement," according to Frank Mankiewicz, Washington-based vice-president of Hill and Knowlton, a major PR firm. "They sense that there is a majority out there and that the emotions are all on the other side—if they can be heard."[30] As a result, environmentalism has become the latest altar of corporate philanthropy, integral to the greening process that has characterized business culture in the past ten years. Public relations maven Jack O'Dwyer describes it bluntly on the front page of his monthly newsletter, which is read by almost every PR practitioner in the country: "Cash-rich companies are funding hard-up environmental groups in the belief that the imprimatur of activists will go a long way in improving their reputation among environmentally aware consumers."[31]

Individual corporate executives have their own rationales. "Because of the type of business we are in we need to prove that we are responsible corporate citizens," explained Chevron's contributions counsel David McMurry in 1992. "Environmental pollution issues are at the forefront in our company, so we are following this up with contributions."[32] That year Chevron donated $1.6 million to environmental causes. ARCO gave almost a million.

Nor is the initiative entirely one sided. Mainstream organizations like the National Wildlife Federation and the World Wildlife Fund openly and aggressively solicit corporate support. A WWF brochure designed for the purpose says that "Your company can use a World Wildlife tie-in to achieve virtually every effort in your market plan. . . . New Product Launches; Corporate Awareness; New Business Contacts; Brand Loyalty." Brand loyalty snagged the Jaguar Car Company. In the same brochure, WWF boasts that Jaguar had committed funds to support the Cockscomb Jaguar Preserve in Belize.

Mankiewicz believes that corporations are supporting environmental organizations for the wrong reasons. "They think the politicians are going to yield to [public] emotions. I think the corporations are wrong about that. I think the companies will have to give in only at insignificant levels. Because the companies are too strong, they are the establishment. The environmentalists are going to have to look like the mob in the square in Romania before they prevail."[33] That, of course, is the last thing the corporate sector wants to see happen.

Since the *Exxon Valdez* spill corporate contributions to environmental organizations have soared to over $20 million a year (about 6 percent of all corporate philanthropy).[34] Although the giving creates a dilemma for some environmental leaders who are wary of the impression it creates, corporate executives have had little trouble finding a safe home for their money in the mainstream environmental movement. By the late 1970s it had become routine practice for corporations to offer environmental organizations small grants to hold conferences or seminars on issues (usually scientific) of particular interest to the company. "At staff meetings a scientist would often report that company A or B had offered us $5,000 to hold a conference on some chemical compound or other," recalls Richard Ayres, one of NRDC's founders. "To which a lawyer would always respond: 'If we're going to sell the name of the organization, surely we can get more than $5,000.' "[35]

As direct mail falters and foundations are unable to fill the deficits in their budgets, national environmental leaders are relaxing proscriptions against accepting corporate money. "Though activists may at first balk at working with corporate America," Jack O'Dwyer advises his PR clientele, "non-profit groups are beginning to realize that private sector cash can increase an organization's clout and bankroll membership building programs."[36] And corporate executives are dining out on the good publicity.

"As environmental compliance officers try to argue convincingly to skeptical regulators that they are doing their best to minimize harm to nature," explains the *Corporate Philanthropy Report,* "corporate giving managers in the same companies are trying to persuade their customers—and their own employees—that they are actually *leaders* in the environmental movement."[37]

In 1991 Waste Management Inc. (since renamed WMX Technologies Inc.), a company that has broken records for EPA fines and violations, donated $1.1 million to environmental organizations: $75,000 to the National Audubon Society (shortly after which WMI President Philip Rooney joined Audubon's board of directors); $45,000 to the National Wildlife Federation (on whose board WMX's current CEO Dean Buntrock now sits); $25,000 to the World Resources Institute; and $100,000 to the World Wildlife Fund. The Environmental Law Institute in Washington, which studies and in many respects influences the course of environmental litigation, accepts grants from Ford, General Electric, several oil companies, and WMX.[38]

Another favorite of corporate philanthropists is Resources for the Future (RFF), a group co-founded in 1952 by the late Fairfield Osborn, head of the New York Zoological Society, who had established the Conservation Foundation four years earlier. An early advocate and supporter of pollution cleanup technology, RFF today describes itself as "an independent organization that conducts research on the development, conservation and use of natural resources and on the quality of the environment." RFF research, often contracted to major think tanks and universities, raised cost-benefit analysis to a near sacrament in government planning. Mandated as federal policy and regulatory procedure by an executive order of Ronald Reagan, cost-benefit became standard procedure under Bush. No environmental protection or standard could be imposed or enforced under the new order, without taking into account the cost of such regulation to manufacturers or others with economic interest in the matter. Cost-benefit was accepted reluctantly by the Clinton administration and is cited frequently in the court and congressional testimony of corporate litigators and lobbyists.

It is not surprising to learn, then, that Resources for the Future is supported by generous contributions from the American Petroleum Institute, Cyanamid Company, Ashland Oil, AT&T, Chevron, Consolidated Edison, Du Pont, Exxon, Georgia Pacific, Johnson and Johnson, the National Agricultural Chemicals Association, Nippon Oil, Pacific Gas and Electric, Potlatch Corporation, Syntex, Texaco, Unilever, Union Carbide, and WMX, among others.

In 1993 the World Wildlife Fund received donations of over $50,000 each from Chevron and Exxon. In reciprocation for their generosity and cooperation several top officers of these corporations were invited to join the boards of WWF and other environmental nationals. Few things make a corporate executive or lawyer look greener than a seat on the board of an environmental organization. A 1990 survey of seven mainstream environmental boards conducted by the *Multinational Monitor* found 67 individuals who also served as chair, CEO, president, consultant, or director of 92 major corporations (35 of them in the Fortune 500). Twenty-four of the directors were associated with corporations that—like Exxon, Monsanto, and Union Carbide—are on the National Wildlife Federation's list of Toxic 500.[39]

The National Audubon Society is an interesting study in confusion over the issue of board membership. The society was founded in the nineteenth century in protest against the corporate slaughter of wild birds for their plumage. A century later Audubon Vice President Robert San George is quite comfortable with corporate representation on his board: "If they see themselves getting into some kind of conflict of interest they will say so," he believes. San George expressed concern that without corporate representation on its board, Audubon would "acquire a reputation of being anti-corporate or shunning the corporate world."[40] And, he adds, the management skills that corporate representatives bring would benefit the organization. Audubon is hardly in danger of acquiring an anti-corporate reputation. Despite the fact that Exxon killed thousands of birds and sea mammals by spilling 11 million gallons of crude oil into Prince William Sound, the Audubon board voted not to support a national boycott against the company.

To Audubon President Peter A. A. Berle the question of whether Waste Management was purchasing a position on the Audubon Society's board in an attempt to direct the society's policy is moot. (Berle, a prominent Wall Street lawyer, was recruited in 1984 by Audubon Chairman Donald O'Brien, a lawyer for the Rockefeller family. O'Brien himself has been a constant focus of controversy within the organization.) "The National Audubon Society has always sought contributions from the corporate world." Berle asserts. "The primary guiding principle has been that contributing companies not expect that Audubon will *in any way* change its public policy positions because of the contributions [Berle's emphasis]. I can assure you there are never any policy strings attached to a gift and the policies of the Society are not affected by a contribution received or promised."[41]

Berle also sees the matter of inviting Philip Rooney to join the board as a moot point.

> Audubon has always had Board members who worked for various corporations. Right now we have members who are connected to various segments of business including paper, insurance, finance, pharmaceuticals, automobiles, newspapers, aerospace and solid waste. In the past we had members who were senior executives with major energy companies. At the same time the Board also has members who are grass root activists, scientists, philanthropists, academics, lawyers and farmers. I believe this diversity gives the Board its strength and the organization an incredible pool of expertise on which it can draw when addressing the environmental issues before us.
>
> It is important to remember that people, not corporations, serve on Audubon's Board; people with knowledge and expertise important to the organization; people who know how to deal with local zoning boards and how to manage our endowment's portfolio, people with expertise in ornithology, geology, government and approaching major foundations for funding. All of them are important and help us consider all possibilities.[42]

But which possibilities prevail? Richard Regan of the Center For Policy Alternatives in Washington recalls an interesting conversation that occurred when he visited Congressman Bill Richardson (D-NM) with an Audubon lobbyist during reauthorization of the Resource Conservation and Recovery Act.

> The Audubon Society representative and I were explaining the position of the environmental community and grassroots citizen network on oil and gas waste to Richardson's staff member, who asked an important economic question. Who would pick up the small amount of accumulated oil and gas wastes at various mom-and-pop stripper well locations tucked away in rural locations across New Mexico and other less populous states if the current oil and gas wastes were redefined as hazardous? The Audubon representative quickly announced with great enthusiasm that Waste Management was uniquely positioned to handle such a [task].

"Was this a WMI representative I was hearing?" Regan asked himself. "Could this expediency explain why WMI was funding Audubon and its oil and gas project?"[43]

It is difficult to know for sure why one corporation or another donates money to environmental causes. Motives no doubt vary from company to company. But "the key," according to an anonymous corporate source quoted in the *Corporate Philanthropy Report* (May 1992), "is that many companies are using their donations to buy time with environmentalists until their environmental reform efforts pay off."

Grassroots organizers make much of the fact that mainstream leaders allow corporate executives, lawyers, and directors on their boards. "The adversary has joined without being defeated," says James Thornton, a former NRDC lawyer now living in Santa Fe. Corporate affiliation on environmental boards is a legitimate concern, but it begs the question of what is actually occurring. Are corporation representatives coopting the organizations? Are they being used as access to money and power by environmentalists? Are they simply sincere environmentalists seeking a place to express their concern? Bill Newsom, a dedicated environmentalist and California appellate court judge in San Francisco, offers a third possibility. "Don't worry about them, they don't do much of anything," he says of the several corporate executives who sit with him on the Environmental Defense Fund board. Newsom finds his corporate colleagues to be environmentally naive and largely uninterested. "Fred [Krupp] tells them what he thinks EDF should be doing and they nod."[44]

That may be true, but they are also there because they are comfortable being there, and Fred Krupp does nothing to make them uncomfortable. This is particularly true when the representative is someone like Amyas Ames, chairman of Kidder Peabody, the Wall Street investment banking firm that underwrites many of the country's largest public utilities. Upon joining the board, he commented that "EDF's very early policy of working in cooperation with business and industry made this an easy decision."[45]

Crossing the Beltway

A winning strategy in any political or social movement requires a presence in the capital. At some point in the history of every great American movement a federal strategy is applied. In no other movement, however, has a federal strategy absorbed such an overwhelming portion of talents and resources as it has in mainstream environmentalism. It was a natural course for a movement steeped in reform and controlled by so many representatives of the power elite.

In 1969 there were two registered environmental lobbyists in the District of Columbia. By 1985 there were 88. Considering the institutional and staff support required to back up that many lobbyists, it was a sizable investment for the few organizations represented. That year Senator William Proxmire (D-WI) called the environmentalists the most effective lobby in Washington. It was taken as a compliment by mainstreamers. To grassroots environmentalists, however, the Washington establishment began to look less like an idealistic social movement and more like a special interest group indistinguishable from the other 2,000 or so lobbies that circle the Hill.

Pragmatic reform is respectable, and compromise, the lifeblood of democratic government, is implicit in a legislative strategy. No group is going to get everything it wants in a clean air or clean water bill when a dozen or so powerful interests are focused on killing it. So the final product will be a compromise for everyone. For a while the compromise tactic worked for the environmental movement, which had little to lose and everything to gain. If the federal government was willing to regulate polluters, compromise made sense, and some ground was gained.

Buoyed by their gains, mainstream environmentalists failed to face up to the limitations of the legislative strategy or the essentially conservative nature of any U.S. Congress, Democrat or Republican. Congress is far more willing to limit than to eliminate, more prone to regulate than to prohibit, more likely to moderate than to forbid the excesses of industrial production. "No matter what the risks," writes Kirkpatrick Sale, "Congress prefers to have some agency like the EPA formulate and monitor 'safe' levels and 'acceptable' amounts rather than ban or discontinue some product or practice understood as dangerous. As a legislative tactic it was often an effective way to win over needed support, particularly from industry-minded lawmakers, and to convince constituents that action was being taken, but as an administrative tool it was usually ineffective and sometimes counterproductive."[46]

As the focus of environmental strategy shifted more and more to the federal government, Washington offices expanded and field offices either shrank or shut down. To contend with the complexity of government, more lawyers and MBAs were hired. The culture of American environmentalism began to change, and not necessarily for the better. In fact, it can be argued that the movement's problems began when the nationals shifted their talent and attention from the mountains, rivers, forests, and communities they were formed to protect. Organizations inspired by the wisdom and ethics of John Muir, Aldo Leopold, and Henry David Thoreau, which had thrived for decades on raw indignation and volunteer energy, became career havens for progressive lawyers, scientists, and lobbyists.

Although the mainstream environmental movement never resembled a cross section of American society, before the migration to Washington there was some noticeable variety among the nationals. One could distinguish a Sierra Club hiker from the National Wildlife Federation hunter, an Audubon birder from an NRDC biochemist. Each organization had its own priorities and its own projects. By the mid-1980s, however, the distinctions had blurred and groups began to look and sound alike. Staffs, direct mail campaigns, rhetoric, and policies—even wardrobes—became homogenized.

Today one inner Beltway environmentalist is indistinguishable from another—and barely distinguishable from any other Washington lobbyist. All are well educated, well dressed, well spoken, and very polite. They might be working at any law firm, congressional office, campaign consultancy, or direct mail operation in the city. Their day-to-day activities are as likely to involve publishing product catalogs, organizing tours, designing direct mail packages, or marketing credit cards as anything directly related to the environment.

A national membership organization with a staff of hundreds, a multimillion-dollar budget, and a dozen or more ongoing projects requires a certain amount of hierarchy, structure, and professional management. An environmental organization is no different from any other large nonprofit. The problem arises when professional managers aspire to top jobs. When boards acquiesce, as they too often do, organizations like the Wilderness Society, Sierra Club, National Wildlife Federation, and Audubon Society—once fueled by passion and run by committed warriors like David Brower (Sierra Club) and Stewart Brandborg (Wilderness Society)—become the domain of list managers, marketing directors, and organizational development specialists. The move to Washington only exacerbated the trend to bureaucracy.

During the high-flying eighties, instead of creating endowments, mainstream organizations created institutions. They continued a 1970s trend toward adding programs and expanding staffs. They spent more effort and resources on developing entrepreneurial and organizational enhancement skills than on environmental issues. The unfortunate end result is a bland, bureaucratic reform movement devoid of passion or charismatic leadership and hell-bent on reform.

What is sad about all this is the frustration and despair of mid-level environmental professionals, many of whom chose their careers because they wanted to fight for the cause. They truly believed that their work in a large, prestigious national organization would be effective and bring obvious results. Many moved to Washington to compete for their jobs, only to find themselves trapped on a treadmill of fundraising, direct mail strategy, and marketing meetings, working long hours in an urban environment far from the direct experience of nature that inspired them in the first place. They knew, of course, that they would be better off financially if they had taken their law and management degrees to the corporate sector. Yet they remain, despite a gnawing feeling that they are fighting the good fight on the wrong battlefield.

3

Fix Becomes Folly

The Bhopal disaster was a bad day at an organic chemical plant.
James Thornton, Positive Futures

The environmental movement has followed the path of most resistance.
Ed Marston, High Country News

On Earth Day 1970 Nobel Laureate George Wald spoke at Harvard University. He gave what now appears to be the most prescient speech of the day. Wald warned that America was perilously close to allowing "anti-pollution to become a new multi-billion dollar business," a huge and very profitable industry that would repair environmental hazards caused by nuclear waste, lead, radon, and asbestos and offer a portion of its profits to the environmental movement. Under Wald's scenario, pollution would "go on merrily in all its present forms," while we "superimpose a new multi-billion dollar anti-pollution industry on top of it. And in these days of conglomerates, it will be the same business. One branch of it will be polluting, the other branch of it anti-polluting."[1]

Wald's prediction has pretty much come true. The environmental cleanup industry—thriving on massive government Superfund contracts—has become a $130-billion industry growing at four times the GDP and is hailed by its corporate owners as "environmental progress."

Several industry leaders, like Browning-Ferris, a hazardous-waste handler, have earned sumptuous profits and become benefactors of the environmental movement.

Wald did not, however, foresee the creation of mutual funds composed wholly of the securities of these enviro-tech companies and sold as "green funds" to environmentally concerned baby boomers. Nor could he have imagined—and would not have been believed if he had—that anti-pollution industry executives would sit on the boards of environmental organizations; or that environmental leaders would become executives of the new environment industry, as William Ruckelshaus did when he became CEO of Browning-Ferris. But no prophet can think of everything.

The same day Wald spoke at Harvard, Adlai Stevenson, III, addressed students and faculty at Western Illinois University. His subject was national pollution control programs, which he criticized as being negative in approach. They are "pollution control" programs not "clean water" or "clean air" programs, he pointed out. The way government environmental protection was structured, damage to air and water had to occur before a state or federal agency could act to control it. That didn't make sense to the state treasurer of Illinois, among the first of his class to see the folly of reform environmentalism.[2]

After they had lobbied successfully for progressive environmental legislation in the 1970s, the trustees and leaders of the national mainstream organizations elected to remain centered in Washington, confident it was the best place from which to oversee enforcement of the laws they had created. Some presence was certainly required in the capital. Legislation intended to reform federal agencies like the Bureau of Land Management and the U.S. Forest Service—which had become subsidiaries of the timber, mining, and livestock industries—was being effectively circumvented. "Multiple use" remained multiple abuse, and public lands remained the commodities factories of the extractive industries.

By the late 1970s and early 1980s, however, environmental antagonists and their lawyers had shifted their locus of attack from the government's Washington offices to state and local courts, legislatures, and regional offices of the federal agencies. The regulatory agencies formed by the triumphant legislation of the early 1970s were assimilating the values of the industries they were created to regulate. The most strategic place to defend the environment was no longer in Washington. It was in the field, at the grassroots of the nation. This was particularly true in the West, where extractive industries were preparing to make another invasion of the public lands, a target that became more feasible upon the election of Ronald Reagan.

While Reagan was running for president the White House Council on Environmental Quality (CEQ) was putting final touches on a report entitled *The Global 2000 Report to the President*. The report, which was finished before his inauguration on January 20, 1981, warned the new president of

> the potential for global problems of alarming proportions by the year 2000. . . . The earth's carrying capacity—the ability of biological systems to provide resources for human needs—is eroding. . . . If present trends continue the world in 2000 will be more crowded, more polluted, less stable ecologically, and more vulnerable to destruction than the world we live in now. Serious stresses involving population resources and the environment are clearly visible ahead. . . . The efforts now underway around the world fall far short of what is needed.

There is no indication that the report was ever shown to the new president, and it was officially ignored. This should have been a clear signal of what was to follow.

Ronald Reagan, more than any president of modern times, demonstrated how easy it is for a recalcitrant leader to ignore not only the advice but also the very statutes of his own government. Although his attempt to subvert the Resource Recovery and Conservation Act (RCRA) and Superfund was thwarted, he managed to stall environmental progress for a full decade. The few improvements made during his administration cannot be claimed as triumphs by the environmental movement, which essentially went limp for eight years. Instead, they were the byproducts of scandals such as "Sewergate," which led to the resignations in disgrace of Secretary of the Interior James Watt and EPA's Anne (Gorsuch) Burford and the imprisonment of Rita Lavelle.

At the heart of the mainstream movement's folly is an abiding faith that legislation backed up by litigation will adequately protect the environmental health of the nation. Grassroots "enviros" harbor no such faith. They have seen that as long as their mainstream colleagues pursue a federal strategy and position themselves as defenders of legislation, very little real protection will filter down to the local communities.

The fortunes of the movement, instead, seem to rise and fall with the public's perception of administrative sincerity. After eight disastrous years with Reagan, George Bush demonstrated how easily a politician could change that perception, simply by declaring himself an "environmental president." Bill Clinton accomplished essentially the same thing by running with Al Gore.

Staying the Course

A long habit of not thinking a thing *wrong* gives it a superficial appearance of being *right*.

Tom Paine, Common Sense

By responding to Ronald Reagan as they did, mainstream environmentalists missed their best and perhaps last opportunity to lead a significant social movement. During the Nixon, Ford, and Carter years there remained a modicum of environmental sincerity in the White House and its agencies. Some important policy gains were made and—while there was an unfortunate tendency on the part of professional environmentalists to assume that policy change meant, *ipso facto,* a better environment—the mechanisms for an improved environment were secured by statutes controlling air and water pollution, protecting endangered species, and regulating toxic transport. With an occasional litigative prod, the agencies formed to enforce the new laws could usually be persuaded to do so.

Environmental legislation was reassuring, but loopholes abounded, and industries soon deployed a battalion of lobbyists to Washington to keep them open. As the sole strategy, the relatively soft legislative/litigative approach could never have become an effective weapon against such antagonists. Still, in the early years of the movement, with moderate government leaders, one could use it as a tool, sometimes compromise a little, and achieve some progress.

But Ronald Reagan was a counterrevolutionary whose policies and tactics quickly revealed the limits of compromise. He was determined from the outset to turn America away from environmentalism. He and his advisors also made it clear that they were ideologically repulsed by compromise on all fronts. They were not the sort of people with whom it is easy to bargain. To the Reagan team social progress meant material abundance and the most important leading indicator was consumption. The Consumer Confidence Index became the vital economic benchmark and was reported dutifully by the media as a harbinger of well-being.

During the 1980 presidential campaign Reagan and his advisors openly refused to meet with citizen-environmentalists, whom they believed to be politically irrelevant.[3] During the transition, even Republican environmental experts and officials from previous administrations were completely ignored. Shortly after he took office, Reagan and other administration spokespersons labeled "environmental elitists" enemies of the conservative revolution—and the state.

James Watt, protégé of the Coors brewing family, champion of the large landowner, friend of the offroad vehicle operator, and a veteran of the Sagebrush Rebellion, variously referred to environmentalists as "Nazis" and "Bolsheviks." Immediately upon taking office as head of the Interior Department, Watt announced his intention to "reverse 25 years of bad resource management." Railing against "the power and self-righteousness of Big Environmentalism," he declared that national stewardship, an obsolete ideal, would be discarded. Watt's admirer and colleague Anne (Gorsuch) Burford was recruited to emasculate the EPA. In her first year in office, EPA enforcement proceedings dropped almost 70 percent.

Meanwhile the president slashed the CEQ budget 62 percent, reduced its staff to 10, and eliminated many of the tests CEQ regularly performed to assess the nation's water and air quality.[4] Reagan also ordered the Office of Management and Budget (OMB) to subject all environmental regulation to rigid cost-benefit analysis and packed the Nuclear Regulatory Commission with proponents of nuclear power. When ozone depletion became a documented threat, Reagan's secretary of energy, Donald Hodel, prescribed darker sunglasses and stronger suntan lotions. War was all but declared on the Sierra Club, the Wilderness Society, and other champions of federally mandated land conservation.

The response of the nationals was verbally loud but politically anemic and set the tone of mainstream American environmentalism for at least a decade. The Sierra Club, which did not publicly oppose the appointment of James Watt to Interior, in 1985 responded strategically to the Reagan presidency by appointing Republican Douglas Wheeler (a former Nixon appointee) executive director of the club.

Watt sought to disrupt the environmental movement by exploiting the persistent conservationist/preservationist divide. Watt believed he could split the "hook-and-bullet boys," whom he perceived as his friends, from the "daisy sniffers" and the movement would collapse. It didn't work. Lifelong Republican conservationists turned against him. Even National Wildlife Federation President Jay Dee Hair, a friend of hunters, fishermen, and Republican presidents, saw in Watt an enemy of the environment.

It was nonetheless difficult for mainstream leaders to respond quickly or appropriately to Reagan's assault. Their organizations had grown so large, inflexible, and inert that they could only stand and watch as the Heritage Foundation mandate for government was implemented and the Task Force for Regulatory Relief under Vice President Bush protected polluters from the EPA and Detroit from emission standards. The reaction to James Watt, however, was immediate and particularly intense, for he was threatening the ultimate sacred cow of mainstream environmentalism—the wilderness. To the conservation nationals the evisceration of EPA was less alarming.

Although Reagan's partisans rekindled some zeal among environmentalists, they inspired no imaginative strategies. Organizing meant still more direct mail campaigns, and activism meant more lobbying. Confrontation took the form of letters—mostly form letters—to representatives and senators. Some congressional opposition was created but not enough to stop Reagan from reversing two decades of environmental reform. To corporate polluters, clearcutters, and strip miners the Reagan decade was heaven sent.

The Group of 10

Environmental CEOs are gearing up to fight last decade's battles with last decade's weapons, on last decades battlefield.
Ed Marston, High Country News

When he saw what was unfolding, Robert Allen, executive director of the Kendall Foundation, convened a discreet meeting of nine mainstream leaders. Allen hoped to spark a powerful coalition to confront head on the clear and present danger he saw looming in the White House. The group met on January 21, 1981, the day after Reagan's inauguration, over dinner at Washington's Iron Grill Inn, a block or two from the White House.[5] Attending were Michael McCloskey of the Sierra Club, Russell Peterson from the Audubon Society, John Adams of the Natural Resources Defense Council, Rafe Pomerance from Friends of the Earth, Louise Dunlap of the Environmental Policy Institute, Jack Lorenz of the Izaak Walton League, William Turnage from the Wilderness Society, Janet Brown of the Environmental Defense Fund, and Thomas Kimball from the National Wildlife Federation (NWF). They were joined by Allen's employers, the brothers John and Henry Kendall, who had sold their family business to Colgate Palmolive and with the proceeds created a foundation that granted about $2.25 million a year to environmental organizations, most of whose leaders were at the table. "Many of them had never met before," recalls Allen, who hoped "that by coordinating their efforts the environmental movement could be rejuvenated and strengthened."[6]

Allen invited to this soirée only officers of organizations that were "active," that is, that regularly met with members of Congress and corporate representatives."[7] This requirement immediately eliminated apolitical organizations like the World Wildlife Fund and the Nature Conservancy and precluded the participation of established organizations like Environmental Action and Greenpeace (the largest of all memberships groups) that distrusted the federal government and defined "active" more broadly. Allen clearly sought to exclude groups conducting, supporting, or advocating direct action against polluters, whalers, the military, and, even more troubling, against corporations.

It was agreed that only CEOs of the organizations selected could attend future meetings. Only one leader expressed reservations about the idea: Jay Hair of the National Wildlife Federation said outright he would not join if Defenders of Wildlife, whom he perceived as anti-hunting, were invited to join. According to Bill Turnage, who was then director of the Wilderness Society, Hair also expressed reservations about the usefulness of collaborative politics.

This didn't surprise Turnage, the most recent arrival in Washington. Before he left California to take over the Wilderness Society, photographer and lifelong environmental activist Ansel Adams, who knew most of the players well, had warned him: "You're about to go work with the biggest egos on the planet. They don't get paid much so the drive is ego, and the righteousness is *self*-righteousness." Adams added, "the worst kind." Turnage says Adams foreboding proved accurate. "I've never see so much territoriality and rivalry. Some rivalry is healthy, but this was counterproductive." There were, however, some good meetings, "although the organizations' staffs disliked each other so immensely that it was hard to get them to collaborate on anything we decided to do together."[8]

At that first meeting concern was expressed about the new administration and its threat to the hard-won legislation of the previous decade. The discussion, Allen recalls, "was surprisingly frank and open. Budgets were revealed, direct mail strategies compared, and consultants' phone numbers exchanged." Former Sierra Club Director Michael Fischer later reminisced about the group's early meetings: "One of the early dividends was the realization that three of the organizations shared the same direct mail consultant—Craver Matthews and Smith."[9] Was this the environmental agenda for America?

Given the state of siege in Washington one might have expected a bold manifesto, even a declaration of war from this awesome force of leaders supported by legions of generous donors and active voters. Instead it was decided not to create any formal organization or to adopt any particular strategy except to meet again in three months with a new chairman, who would determine an agenda. At the next meeting, as if to minimize the threat of their existence, the 10 who gathered decided to close ranks and, though reluctant to give themselves a name, to call themselves The Group of 10. For a decade or more the G-10 became synonymous with mainstream environmentalism and represented what the American media meant when they referred to the environmental movement.

During the Reagan years the original G-10 organizations more than doubled their memberships and staffs. The combined memberships of the nationals rose from around 4 to over 7 million. The Wilderness Society, for example, grew 144 percent between 1980 and 1983 and from 68,000 in 1981 to 350,000 members today. The Sierra Club increased its membership by 90 percent during the early eighties, Defenders of Wildlife and Friends of the Earth by about 40 percent.[10] Toward the end of the decade most mainstream groups experienced additional surges in membership, which were stimulated by reports of ozone destruction, global warming, images of oil on the beaches of Alaska, and medical wastes on the beaches on New Jersey.

The Wilderness Society's budget expanded more than tenfold in the decade from $1.5 to $16 million, while the Sierra Club and the National Wildlife Federation experienced similar financial growth. By the end of the 1980s all but four of the Group of 10 had moved their headquarters to the capital, and two of the others were considering the move. What originated as a special interest and evolved into an active social movement began to look more like a trade association or a cartel.

G-10 leaders met quarterly for a few years, conducted occasional joint projects (usually research), and formed a "B-team" of their conservation directors and top lobbyists. According to participants willing to discuss the subject, "there were some very productive meetings" and "some that went nowhere."

The first visible product of the group's work was a 1986 report published as *An Environmental Agenda for the Future*. The report reflected the consensus of the members, who described themselves boldly as "leaders of America's foremost environmental organizations." Creators of the *Agenda* were clearly convinced that human overpopulation was "the root cause" of environmental problems. The implied solution to environmental degradation was "formal population policies," good science, and resource management. Pollution was seen as a technological rather than a political challenge. Environmentally controversial subjects like nuclear energy and the petrochemical industry were completely avoided. "It was a log-rolling exercise," recalls B-Team leader Richard Ayres, a founder of NRDC and the organization's first lobbyist. "The leaders were simply bowing to each other's agendas. They never developed a coherent overall direction or strategy for the movement."[11]

While acknowledging that existing laws and regulations did not address the root problems, the *Agenda* failed, with the single exception of population control, to offer any policies to confront environmental degradation head on. Environmental CEOs, it seemed, were unwilling to sacrifice their few comfortable relationships in the Reagan administration or take a stance that would isolate them from their friends on the Hill and in industry. The report offered no critique of America's petrochemical dependency or its disastrous transportation policy and made no strong recommendations for renewable energy sources, sustainable-yield logging, or mass transit. In all those categories it encouraged only more research. The response to chemical-based agriculture was, again, to encourage research into organic-farming methods. It goes without saying that the report shied away from advocating any further public control of production technology. The leaders of the Group of Ten had obviously chosen to play by contemporary Republican, District of Columbia rules. If the report was any indication, Robert Allen's bold initiative had led to little more than a reinforcement of reform environmentalism's worst tendencies—indiscriminate compromise and capitulation to entrenched interests.

Grassroots criticisms of the *Agenda* were swift and at times blistering. "In their reflexive adherence to legislative solutions," wrote Colorado editor Ed Marston, "they sound like chemical companies responding to the increased resistance of pests to pesticides by prescribing even larger doses of ever more potent poisons. The environmental movement," Marston continued, "has taken the path of most resistance by concentrating on the Congress and the courts to accomplish its end . . . As *Agenda* demonstrates, the national groups appear unable to look in any direction but straight back."[12]

Other responses to the administration from the G-10 occasionally took the form of tough rhetoric from the sidelines. More hard-hitting direct mail copy was written to exploit the Watt factor. In the early Reagan years, over 100 million prospect letters were mailed (virtually none of them printed on recycled paper). Mainstream support grew exponentially as millions of dollars poured into environmental coffers.

How wisely was it spent? The expansion of the movement was largely horizontal. More lobbyists and lawyers were hired. More amendments were drafted and more lawsuits filed. But far-reaching campaigns were weak, and on the battlefield little changed. The legislative/litigative strategy remained central. Initiatives made by the boards of mainstream environmental organizations tended to involve shifts toward accommodating the free market ideology of the new administration—an administration whose agencies and departments either ignored, watered down, or broke outright the laws they were mandated to enforce.

All of this occurred while national polls were discovering that Ronald Reagan's environmental stand was out of line with public opinion. All his efforts had not lowered the public's commitment to environmental protection. Throughout his administration polls showed increasing support for environmental regulation and even a willingness to sacrifice economic growth for additional protection.[13] By 1988 over 90 percent of Americans were comfortable calling themselves environmentalists, and over 60 percent approved of a tougher federal role in environmental regulation—even if it required additional taxation. If there was ever a time in the history of the environmental movement to initiate aggressive litigation against intransigent regulators, conduct mass demonstrations, organize consumer boycotts, file shareholder suits, and impose nonviolent direct action against polluters the 1980s was it.

Only grassroots environmentalists seemed nimble enough to address new opportunities and did so as well as they could with their limited funds. Meanwhile mainstreamers had acquired an addiction they couldn't kick—access to power, a powerful inebriant suffered by many in Washington. Symptoms, sometimes identified as "Potomac fever," include abiding trust in legislated mandates, faith in effectiveness of the lobby, and reliance on ephemeral voting blocks to mitigate social and political problems. Throughout the Reagan years, the reform environmental movement displayed all the above symptoms. And, like any addict with enough money to support a habit, they either denied the problem or explained it away.

Although environmental leaders became disillusioned with government (exactly what Ronald Reagan hoped would happen), most chose to remain on the comfortable reformist path. Thus the very system they had contrived to reverse the degradation of the environment began to fail them. It might have failed even without the Reagan assault, for no strategy works forever, but against such hostilities accommodation could never have succeeded. It could even be argued that President Reagan offered the mainstream organizations their last opportunity to become a powerful and effective force and that they passed it up.

At exactly the moment it should have moved vertically, the movement went horizontal. It played safe. Instead of reaching out to state, local, and regional grassroots organizations, it formed an exclusive Beltway club of white and (all-but-two) male CEOs. It even excluded its own lobbyists and political directors from strategic deliberations. Had the impetus been vertical, the American environmental movement today would be much stronger and more consequential than it is. An explosive critical mass of national activism could have been formed. Instead, a relatively harmless and effete new club appeared. The principal achievement of the G-10 was to give environmental leaders a forum from which to quell some of the conflicts that inevitably arise among organizations in the same movement. It could have accomplished so much more.

Almost 10 years after its first meeting the G-10 opened its doors to additional members. Renamed the Green Group, the original 10 were joined by the Union of Concerned Scientists, Zero Population Growth, the World Resources Institute, and (perhaps most promising of all) the Children's Defense Fund. Inclusion of the latter at least hinted at the badly needed nexus between environmentalism and human rights.

Even so, the coalition remained limited. When emissaries from the new group went to invite Greenpeace—by then the largest environmental organization in the world—its director Peter Bahouth politely asked if Lois Gibbs would also be invited. Gibbs, the feisty mother from Love Canal, had founded a large national organization called the Citizens' Clearing House for Hazardous Wastes and had a reputation for attacking mainstream environmental organizations that compromised with waste handlers and undercut her work at the grassroots (see Chapter 6). Mike Clark, former director of Friends of the Earth, admitted that Gibbs was not being invited. "When she is," Bahouth replied, "Greenpeace will come to meetings."[14]

The results of the expansion have not been entirely positive. Some of the original leaders, most notably Jay Hair, have expressed frustration and stopped coming to meetings. The amity that once existed among mainstream leaders, at least the heads of the 10 largest organizations, seems to have evaporated. Disagreements about the North American Free Trade Agreement (NAFTA) (discussed in Chapter 7) were partly to blame. The inclusion of less accommodating enviros like Jane Perkins, former executive director of Friends of the Earth, and Barbara Dudley, Peter Bahouth's successor at Greenpeace, has probably been more disturbing to conservative members.

In a private 1994 memorandum to the NRDC's John Adams, former Sierra Club Executive Director Michael Fischer expressed concern that the Green Group was losing effectiveness. He believed this was partly because "the economy of the 1990s has exacerbated the normal competition among organizations." The environmental movement, Fischer wrote, was beginning to look like "a zero sum game [which has made] the sharing of agendas, marketing approaches and statistics about direct mail approaches problematic." The Group of 10, which had once been a collegial gathering of like-minded men, had lost its "clubbiness" (Fischer's word), become almost ineffectual, and been branded by outsiders as "the Gang of Ten."[15]

Fischer, who now works outside the movement as director of the California Coastal Conservancy, proposed a few possible options for the future of the Green Group. Why hold meetings at all, he asked Adams. His own answer was "to assure peace within the community . . . and search for collaborative opportunities." He suggested meeting less frequently and breaking into smaller groups that reflect special environmental interests. His strongest recommendation, however, was to hire a "secretariat" to coordinate the activities of the Green Group and help environmental "CEOs set priorities, set joint action plans, enhance professional activities . . . develop strong positive working relationships and define and articulate a unified vision for the environmental movement."[16]

The Green Group still meets, usually in Washington, occasionally in New York. However, the agenda is federal, the battleground is Capitol Hill, and the strategy is about as imaginative as ever.

Going Along to Get Along

We're not selling out, we're buying in.
Jay Dee Hair, National Wildlife Federation

Let the people we pay to compromise—the legislature—do the compromising. . . . Every time I compromise I lose.
David Brower, For Earth's Sake

As I indicated in the preceding chapter, habitual compromise is not new to American environmentalism. Nor has it gone unchallenged, even by mainstream enviros. One who experienced the hazards of compromise early in his career was David Brower, mountain climber, activist, and former executive director of the Sierra Club. Brower, who is founder of Friends of the Earth and the Earth Island Institute and co-founder of the League of Conservation Voters, is regarded as the gray eminence and "arch-druid" of American conservation.

In the 1950s, when he was running the Sierra Club, Brower and his friend Howard Zahniser, at the time CEO at the Wilderness Society, cut a deal with the Bureau of Reclamation. They agreed not to oppose construction of a dam in Glen Canyon in return for the bureau's promise to cancel two dams that would have flooded Dinosaur National Monument. The bureau agreed and both sides claimed victory. The Bureau of Reclamation celebrated its new lake in Glen Canyon—" The Jewel of the Colorado"—and the Sierra Club declared Dinosaur a triumph of environmental activism.

In mid-June of 1963, after he had celebrated the agreement but before the dam was finished, Brower floated through Glen Canyon in a boat and saw what he had traded away. "It was not easy to travel the distance without reverence, or without being grateful that an already beautiful world could here exceed itself," he wrote in sadness. "In this cathedral, whether you looked at gardens, at the altar or the stream that flowed by the nave, you knew what this place meant to its setting. There would never be anything like it again."[17] After the trip Brower swore to himself and the public that he would never compromise again. Unfortunately he did, one last time, over the Central Arizona/Grand Canyon dam fight a few years later. The project led to expansion of coal-fired power plants throughout the Four Corners area of the southwest and became an air pollution nightmare.

Brower still cautions his fellow activists to eschew compromise. "The problem with too many environmentalists today is that they are trying to write the compromise instead of letting those we pay to compromise to do it," Brower says. "They think they get power by taking people to lunch, or being taken to lunch, when in reality they are just being taken. They don't seem to learn, as I did at Glen Canyon, that whenever they compromise they lose."

The advantage to compromisers is that when a compromise position prevails, as it almost always does in politics, the compromisers on both sides of an issue can declare victory, boast of their triumph in mailings or campaign speeches, and (most important of all) be invited to the next compromise conference (that is, they gain "access").

Compromise, however, is not always an appropriate strategy. It works best when it comes from a synthesis of extremes. If one party comes to the table from a centrist position or with a reputation for accepting centrist compromises the settlement will probably lean in favor of the other party. If, for example, rapid, high-profit, heavy-polluting mega-technology is one extreme, and some acceptable level of human and environmental health is the compromise position, what should be the other extreme? And who can best represent it, a centrist or an "extremist"? If, as many argue, accepting compromise is realistic politics, the question remains as to what role environmentalists and environmental organizations play in that process. Should they be the compromisers and come to the table as such? Should they be uncompromising fighters who represent only the environment? What it comes down to is whether environmental advocacy organizations should be advocates or mediators.

When government and the environmental movement both play mediatory roles and compromise on environmental issues, it is left to "radicals" to create the extreme position to counter the mega-technological position, be it preservation of pristine wilderness or a toxic-free world. Between two such extremes some degree of environmental protection can, hopefully, be brokered through compromise. But when radicals are excluded from the process, as they are by both government and mainstream environmentalists, there is no extreme against which to negotiate. Government, polluters, and environmentalists are then negotiating in relative harmony. The result is scant progress.

The lesson of Glen Canyon, and of scores of other environmental defeats, is that compromise will only gain results if one party is willing to do something more drastic than cut a deal. Unless some enviros at the table are prepared to walk out, sit in at a government office, picket a polluting factory, boycott a product until it hurts, or even chain themselves to a thousand-year-old redwood, attempts to compromise will lead inevitably to failure.

Litigation Redux

My primary emotion in recalling the past 20 years of environmental law is one of profound disappointment.
Rick Sutherland, Earth Island Journal[18]

Nothing has served conservation so well over the past 25 years as litigation.
Vic Sher, Sierra Club Legal Defense Fund

Environmental litigation is nearing the end of its useful life.
Joe Bodanich

Operating in Washington forces any organization in any social or political movement to think in terms of legislative strategy and, eventually, to become a lobby. Following that strategy means that litigation will be necessary to enforce and protect legislative gains. Thus the birth of the environmental law firms.

As the previous chapter demonstrates, "suing the bastards" worked best in the early days of the environmental movement when the courts were open to the idea of enforcing regulatory statutes and judges were sympathetic to environmental concerns. The bench, alas, has changed radically over the past 15 years. Today it is dominated by conservative jurists, and judicial conservatism has become infectious, spreading even to the last remnants of the liberal bench. Taking advantage of the situation, the anti-environmentalists who now employ the preponderance of environmental lawyers in the country have abandoned their defensive position and taken the offensive, litigating aggressively for regulatory relief and private property rights. Traditional environmental lawyers, the genuine advocates for the environment, are forced into a defensive stance. "We taught them well," says a somewhat embittered Sharon Duggan, a San Francisco-based private practitioner who specializes in timber sale litigation. "Now they proceed against us."

Gradually, adherence to the three Ls—lobbying, legislation, and litigation—has eroded the potential of other strategies, by robbing them of talent and resources. As antagonists have developed better counterattacks, the effectiveness and power of the entire environmental movement has diminished as well. And, worse yet, the citizens whose vital interests are at stake have been moved another step away from the process. "In public interest lawsuits, the inevitable bargaining over final decisions often gets left to the capable few who have the capacity to undertake the work, especially those people with law degrees," explains William Greider. "Whether virtuous or otherwise, these agents bring their own particular values to the table and their own class biases, which may or may not harmonize with the larger public that cannot be present and has lost reliable representation."[19]

Then there is the matter of the bench. By the end of George Bush's presidency, Bush and Reagan between them had appointed more than half of the 714 federal district, appellate, and supreme court justices in the country. Most of these appointees are hostile to federal regulation and to the very notion that private citizens or their organizations should be able to redress policy grievances through the courts.[20] Although some conservatives turn out to be more conservationist than Ronald Reagan might have hoped, many of the most environmentally hostile jurists will be seated for a long time to come. Judges are throwing out environmental and safety regulations at every opportunity and ruling against environmental organizations in case after case that would have been won easily a decade ago. Once a friendly citadel of environmental protection, the U.S. Courts in every district of the nation are insisting that health and safety regulations be cost-efficient; are challenging federal agencies' authority to impose regulations, and ruling that plaintiffs that once had standing in their court no longer do so.

In March of 1992 alone the administration won cases in federal court that relieved the automobile industry from installing pollution devices, blocked a Labor Department proposal to reduce workplace exposure to toxic chemicals, and sided with the coal industry in a suit that would have limited western strip mining. "The courts traditionally have been the last refuge where we could seek relief from environmental pollution and for health protection," Albert Meyerhoff, senior attorney for NRDC told *The New York Times* in March of 1992. "More and more we are seeing that the courts are unavailable to us."[21]

Even President Bush's own EPA Director William Reilly was distressed by a ruling that overturned an EPA ban on asbestos. The court said the agency had failed to consider the safety consequences of removing asbestos from brake linings and criticized the cost of the ban (calculated as $2.2 to $5.4 million per case of cancer). "The decision to overturn a rule based on ten years of careful research and analysis is at odds with my concept of judicial restraint," Reilly told *The New York Times*.[22]

David Doniger, a former lawyer with NRDC, recalls the early eighties, when the federal bench was still presided over by Kennedy and Johnson appointees. "We found the courts to be very receptive then. You could almost go in with a calendar rather than a brief and say: 'The law says this is supposed to happen in six months or a year,' and you'd get a court order. More recently we've had more trouble. The appeals court in the District of Columbia is being more aggressive in saying, if there is no explicit deadline, we're going to assume Congress did not intend one and we'll give the agency as much time as it damn well pleases."[23]

Departing from their normal domain as legal interpreters, Reagan/Bush appointees have even made judgments about the cost-benefit of specific regulations. "Courthouse No Longer Environmentalists' Citadel" read a March 23, 1992, headline of a *New York Times* article summarizing the loss of several landmark environmental cases. EPA Administrator Reilly argued that the courts were "second guessing" regulations that protect environment and health. "If this second guessing becomes a trend in judicial decisions, the upshot may well be to reinforce Congress's temptation to prohibit any calculation of cost, no matter how small the risk or how great the expense," Reilly said.[24]

As has happened in so many other branches of government, the federal courts have gradually been won over by the regulated industries.

New Federal Strategy?

In order to accomplish anything in Washington, groups have to participate in the system and have access to policy makers.

William Reilly, EPA Administrator in the Bush Administration

With the federal bench loaded with so many conservatives appointed for life, 1992 might have been a propitious time for the environmental legal community to shift strategies in favor of more state and local litigation—if indeed litigation was the best environmental strategy. Once again, inertia prevailed. The environmental law firms persisting focus on a federal strategy made it difficult to win battles at the local level.

"Adopting national air ambience standards in the Clean Air Act was the biggest mistake we ever made," according to Leon Billings, an environmental consultant and former staff director of the Senate Public Works Committee. "Citizens-for-clean-air groups were beginning to get local governments to adopt tougher standards around the country, much tougher than anticipated, and industry wanted to get out from under these local and regional activists who were coming after them everywhere. The Clean Air Act brought the fight to Washington where industry could manipulate things much more cleverly."[25] And the nationals reflexively joined the process. When the bill was passed, they declared victory. But was it a victory?

Not according to Billings. "The federal law short-circuited the activism." he asserts. "It took away the forum for local activists and they had to become involved in much more technical arguments, an arena where industry is strong and citizens are weak. Once policy issues become engulfed in the federal bureaucracy, the public loses the ability to influence those decisions because local electronic and print media simply lose interest in the issues. The story is suddenly distant and difficult to cover in a local newspaper. There is no local agitation because it is now a 'national issue.' "[26] Once again national organizations were complicitous in the weakening of grassroots power.

A more recent problem with federal legislation concerns the funding of environmental mandates. Because the government runs such an enormous deficit it has become difficult, if not impossible, to appropriate funds for legislation. Regulatory bills, therefore, are drafted without appropriations, giving fiscal conservatives less reason to vote against them. At state and local levels the practice has created a rebellion against unfunded mandates, which burden state and local budgets with the cost of enforcing such things as water and air quality standards.

The backlash has led to the introduction of "no money, no mandate" bills, which would eliminate all federal rules affecting state and local government that fail to provide federal funds to carry them out. Such initiatives, usually introduced in the form of amendments to reauthorizations of environmental legislation, are being sponsored by majorities in both the House and Senate. The *reductio ad absurdum* of funded mandates would force the federal government to support all municipal water systems, make grants to auto companies to develop and install airbags, and pay dairies to pasteurize milk. The absurdity of that logic has not impeded, of course, the increase of antagonism toward environmental legislation.

4

Antagonists

Increasingly we are all environmentalists. The President is an environmentalist. Republicans and Democrats are environmentalists. Jane Fonda and the National Association of Manufacturers, Magic Johnson and Danny DeVito, Candace Bergen and The Golden Girls, Bugs Bunny and the cast of Cheers are all environmentalists.

Richard Darman, Director, Office of Management and Budget, Bush Administration

Americans did not fight and win the wars of the twentieth century to keep the world safe for green vegetables.

Richard Darman

Even the most polite and accommodating political movements create their own adversaries, small and large antagonists that form inexorably, like killer T-lymphocytes in the body politic, to combat each threat of change. The problem with the lymphocyte metaphor is that killer T-cells are a sign of good health. Environmental antagonists are not. Yet they are just as inexorable, and as stealthy.

All movements breed such antitheses, as Hegel called them—ideas and tendencies generated by opposite ideas and tendencies. Few, however, have stimulated such virulent antagonism against themselves as American environmentalism, which, by its very nature, threatens the most sacred institution of our culture—private property. The movement's aggressive litigation has aggravated powerful and corrupt government bureaucracies and harassed the most powerful economic institution ever formed, the transnational corporations. By attacking government and corporate sectors at the same time it stimulated a well-organized, well-financed backlash. After the election of Ronald Reagan foundations funded by conservative business leaders like Joseph Coors and Richard Mellon Scaife mounted an ideological jihad against environmentalism. It was led by the Washington-based Heritage Foundation, with its $10-million-annual budget, which has found ways to blame environmentalists for almost every social problem plaguing the country.

The handling of adversaries is perhaps the most difficult of all political skills to learn and is a significant test of a movement's effectiveness. The very appearance of powerful antagonists is a sign of success, but there can be no true victory until enemies are vanquished, or at least kept at bay. Therein lies the threat of failure for American environmentalism. Having produced impressive results in its early years, it now seems unable, or unwilling, to confront the adversaries it has created.

In a few cases environmentalists have cleverly exploited some antagonists like James Watt and Ronald Reagan to expand their constituency. Others, like labor and the Wise Use initiative, they have badly misread. Foreseeing antagonism and keeping it at a disadvantage can strengthen and enlarge a movement, just as periodic illnesses can strengthen the immune system. Attacks from the Heritage Foundation, the Sagebrush Rebellion, the Christian Right, and numerous associations formed by the extractive and petrochemical industries have created opportunities for growth in the American environmental movement. During the 1980s the movement did grow, largely in response to these assaults. But it grew only in numbers.

Today a fractious and weakened movement faces the most aggressive adversary in its history. Operating nationwide, copying many of its own well-tested methods, and borrowing its very name from a central tenet of turn-of-the-century conservation, the wisely named Wise Use movement has appeared in the American body politic like an elusive virus, hard to fathom and very easy to misread.

The Industrial Lobby

Total pollution elimination is not an optimal solution.
Christopher Boerner, Center for the Study of American Business

When *Silent Spring* first appeared as a three-part serial in the *New Yorker* magazine, before publication of the book, Velsicol Chemical Company, a major manufacturer of pesticides, attempted to suppress it by filing a lawsuit against the publisher, Houghton Mifflin. Rachel Carson, they argued, was creating "the false impression that all business is grasping and immoral." *Silent Spring,* they added, intended to "reduce the use of agricultural chemicals in this country and in the countries of western Europe so that our supply of food will be reduced to east-curtain parity."[1] The book was published, but Velsicol's attempted injunction, and the chemical industry's vicious $250,000 public relations attack, foretold the tone and strategy of a new antagonist set upon fighting the "ignorance," "bias," "hysteria," and "paranoid fears" of a powerful new environmental movement. In the years that followed, the industrial lobby grew exponentially in size and sophistication.[2]

No matter how large, clever, and sophisticated in the ways of Washington the environmental movement has become, when it comes to lobbying Congress, it has remained a mosquito on the hindquarters of the industrial elephant. Corporations finance a lobby that is willing to spend almost unlimited time and money combating a process—environmental regulation—they claim costs them $125 billion a year. Chemical manufacturers, oil companies, big agriculture, timber interests, and their PACs will, unless campaign finance laws are reformed, always have greater access to the legislature than environmental lobbyists. This is a reality that Washington-based environmental organizations have had great difficulty accepting.

Even before they contribute a cent to their PACs the companies invest in public relations and media campaigns an amount surpassing the entire lobbying budget of the Green Group organizations. The corporate anti-environmental PR outlay, estimated at $500 million for 1990, buys campaigns using many of the same tactics environmentalists use against them. The most notable are direct mailings from organizations with names like the National Wetlands Coalition (oil and gas industry), the U.S. Council for Energy Awareness (nuclear industry), and the Alliance for America (timber, mining, and cattle). It also finances media buys for "public interest" op-ed ads like the Mobil series and Chevron's greenwashing "People Do" advertising campaign.

Studies of lobbies and PAC contributions indicate that industry is pretty much willing to match the environmental movement about 10 to 1 in dollars and lobbyists.[3] In the 1991–1992 congressional session, the Sierra Club contributed $680,000 to congressional candidates nationwide, an enormous amount for an environmental organization. The amount was dwarfed, however, by the $21.3 million donated during the same session by the energy and resource-extraction industries alone.

So why would the club squander $680,000 against all those millions? The answer, according to Sierra Club Political Director Daniel J. Weiss, is "access". "The more you can talk to legislators, the more you can influence them, and from that influence comes policy decisions." But how big a foot-in-the-door can Weiss buy? During the reauthorization of the Clean Air Act in 1990, he was granted 30 minutes with a key member of Congress. A General Motors vice president spent four hours the very next day with the same representative. "Buying influence," Weiss concludes, "is an arms race that the environmental community cannot win"—a telling statement from an active practitioner of the art.[4]

"If I represent an industry, I can always get into the argument in the Executive Branch or the Congress by nature of the fact that I have money," explains Curtis Moore, former Republican counsel to the Senate Environmental Affairs Committee. "But if you're an environmentalist, you can't get into the argument unless they want to let you in. And they're not going to let you in if they think you are crazy, if you don't think in the same terms they do. You have to sound reasonable or you don't even get in the room. So you don't find many people in major environmental groups who are willing to be seen as unreasonable."[5] And in Washington people with money do seem more reasonable than those without it. Yet the folly continues, as puny environmental mosquitoes attempt to compete with the industrial elephant.

The corporate lobby's approach to environmentalism is described by some observers as a "three bites of the apple strategy." The first bite is to lobby against any legislation that restricts production; the second is to weaken any legislation that cannot be defeated; and the third, and most commonly applied tactic, is to end run or subvert the implementation of environmental regulations.[6] The Clean Air Act, which is seriously wounded every time it returns to Capitol Hill for amendments, is but one of many victims of these tactics. The original Clean Air Act of 1970 set a goal of a 90-percent reduction of carbon monoxide, hydrocarbons, and ozone over major U.S. urban areas by 1977. By that year reductions in most cities had barely reached 30 percent. In successively amended

Clean Air Acts, municipalities and their polluters were granted extensions, first to 1982, then to 1987—without serious resistance from mainstream environmental leaders. Recently the three most polluted cities in the country—New York, Los Angeles, and Houston—received additional 20-year extensions.

Epidemiological studies have shown that the citizens of central Los Angeles are killed in large numbers by their own air. Yet in the 1990 round of amendments to the Clean Air Act, which combined two separate air pollution control bills, citizen access to appointed air quality control boards was reduced almost to zero. The Clean Air Coalition, which included representatives from NRDC, NWF, the Sierra Club, the Environmental Defense Fund, and the National Audubon Society, not only participated in the drafting and revision of the new bill but validated the use of "risk assessment," "acceptable risk," marketable "pollution credits," and inclusion of costs in determination of the "best technology." Twenty years earlier, when the original Clean Air bill was crafted, no environmentalist would have countenanced such concepts. Now mainstream enviros tout their participation in that compromise as a victory for the movement, ignoring the fact that they were seriously outnumbered, outgunned, and outmanipulated by polluters and their lobbyists. And that they were hoodwinked by the White House.

In June 1989, as amendments to the bill were being drafted and hearings planned for the fall, President Bush invited members of the Clean Air Coalition to the East Room to meet with him and "executives of some of the most important companies and business organizations in America." There, he wrote, "we can break the stalemate that has hindered progress on clean air. . . . New solutions are close at hand. . . . This can be known as the year we mobilize leadership both public and private to make environmental protection a growth industry. . . . Ours is a rare opportunity to reverse the errors of this generation in the service of the next." That's the kind of language that gets Washington environmentalists excited. Even the media was impressed. "The Year for Clean Air," exalted a *New York Times* headline. The *Wall Street Journal*'s read "Bush, Resolving Clash in Campaign Promises, Tilts to Environment."

Bush wasn't tilting to the environment, he was embracing the Clean Air Coalition, which he knew very well was dominated by industry and environmental representatives content to fiddle with regulatory minutia and limit public participation. The result, he knew, would be a bill that ever so gradually phased out ozone depleters, validated risk assessment, banned no chemicals, controlled no production processes, and preserved cost analysis in all regulatory decisions.

Bush was right. The Senate and House bills were virtually indistinguishable in their willingness to allow continued production of air polluting toxins, grant exemptions and loop holes to the most egregious polluters, and ignore the very basics of atmospheric science. Ultimately the combined bill contained most of the provisions desired by industry and created the notorious pollution credit—all with the acquiescence of the Clean Air Coalition. (See Chapter 5 for more on pollution credits.) Most of the environmental representatives on the coalition now consider the Clean Air Act of 1990 a triumph for the environment and a victory for the movement. And most of their financial supporters, either ignorant or forgetful of history, accept their analysis.

Even without representation on coalitions, whatever the legislation, industry will be stalking the halls of Congress with amendments and have no trouble finding an indebted legislator to attach a rider, fight for an exemption, or simply vote against the bill. That doesn't mean that environmentalists should not be pursue a legislative strategy. It simply means that its potential is limited. As the opposition refines its defenses, the selection of targets becomes more critical, and alternative, more combative strategies need to be considered. This is particularly true when so much ground can be lost so easily by adding a few short amendments reauthorizing legislation—a threat that loomed so large at the close of the 103rd Congress in 1994 that many environmentalists came to believe a temporary moratorium on all environmental legislation might prevent further erosion of environmental standards.

American social history clearly demonstrates that no major advances for rights or liberation have been made, or progress sustained, by legislation alone. In fact, passage of a landmark bill was often but the culmination of a prairie fire lit by abuse or injustice and fanned by massive public protest and direct action into real progress. Continued reliance on the legislative fix clearly illustrates the inertia and lack of imagination that could plague the mainstream movement well into the next century.

Superfund

Few episodes in American environmental history better illustrate the power of the industrial lobby than the protracted response to the awkwardly named Comprehensive Environmental Response, Compensation and Liability Act, a law designed to create a fund for toxic cleanup so massive that it earned the nickname "Superfund." Enacted in 1980 and signed into law by Jimmy Carter, Superfund placed financial responsibility for cleaning up the toxic miasma left by a hundred years of unrestrained industrial expansion on the companies that made the mess. The presidential transition team set their sights on Superfund the first day after Reagan's election, but not even the most somnolent president in American history could ignore the single most motivating statistic: there were over 30,000 abandoned toxic-waste sites in the United States, an average of over 600 per state in the lower 48.

Passage of Superfund was a major defeat for corporate America, and—but for the immediate and very strategic creation of the Superfund Coalition—it would have been a major triumph for the environment movement. The coalition was founded in 1987 by the corporations that would bear most of the financial responsibility for cleanup: Dow, DuPont, Monsanto, General Electric, Union Carbide, and their insurance companies, Hartford, Cigna, Aetna, and others. Its objective was to influence public opinion against the legislation in the years before its renewal. It would attempt to convince the public and the legislature that costs had to be shared by taxpayers if the cleanup was to be carried out. Achieving this goal required "independent" research on the economic impact of Superfund conducted by an expanded coalition with some semblance of impartiality. Environmentalists were invited to join.

To their credit the Environmental Defense Fund, Sierra Club, Natural Resources Defense Council, and Audubon Society all declined. But the Conservation Foundation, then headed by William Reilly, not only joined but agreed to conduct a study of the Superfund law jointly funded by the coalition and the Environmental Protection Administration. That was all the coalition really needed—one reputable environmental organization with its name on a report saying that Superfund, as written, was ill-conceived and unworkable.

The Conservation Foundation (CF) was founded in 1948 by Henry Fairfield Osborn, retired director of the American Museum of Natural History, and funded by Laurance Rockefeller. It was a non-advocacy organization that sponsored conferences and publications on government and corporate resource management and, holding as it did to Osborn's faith in free enterprise to correct its own systemic abuses, a safe choice for the coalition.

In 1988 Reilly accepted for CF a $2.5-million contract from EPA. The next year he was appointed EPA administrator by President George Bush. The study continued at the Conservation Foundation. By 1991 Superfund was under attack in the press. "Experts Question Staggering Costs of Toxic Cleanups" read *The New York Times* on September 1, 1991, just as Congress was beginning to review Superfund legislation. Reporter Peter Passell's story relied heavily on corporate sources from the Superfund Coalition; nonetheless, it seems unlikely that the *Times* would have run such a story and headline without the independent imprimatur of a well-respected environmental organization.

Sagebrush Rebels

You can't just let nature run wild.
Alaska Governor Walter Hickel

Anyone seeking to understand the depth of the hostility toward environmentalism in the American West need look no farther than the wide open spaces of Nevada. Over the last century the cattlemen and miners there have become so accustomed to the virtually free and untrammeled use of public lands that they now see them as their own domain. Any threat of restriction or increase in grazing or mining fees on the 48 million acres owned and managed by the federal government is seen as a infringement of basic rights, a threat to the livelihood of the rugged families whose grandparents and great-grandparents settled and tamed the state.

In 1979, Statute 633 was introduced in the Nevada legislature. It claimed all the land managed by the Bureau of Land Management (BLM) for the people of the state of Nevada. Although the statute passed, it was regarded by the federal government as little more than a "sentiment" bill and was not tested in the courts.

In the 1980s, hostility to the BLM, the National Park Service, and other agencies of the government grew throughout the West and Alaska to the point of rebellion. The *Washington Post* dubbed the periodic outbursts of anger the Sagebrush Rebellion. Large landowners and leaseholders throughout the West decided that the only way to wrest control of the land and its resources was to deed federal land over to the states and then have the states sell it off (cheaply of course) to private interests. The Sagebrush Rebellion combined the cultural icons of cowboy independence, free market ideology, and private property as a sacred human right. Among the major points of rebellion were the right to use herbicides on grazing land and the right to kill raptors, wolves, and grizzly bears at will.

In the local press and at angry meetings government officials were attacked and environmentalists branded as un-American. The rhetorical hostility in western cattle and mining towns, sometimes accompanied by threats and petty vandalism, grew so intense that BLM employees were ordered to travel in pairs. The Nevada initiative spread to Utah, Colorado, Wyoming, and eventually to Alaska, where it became the Tundra Rebellion. Senator Orrin Hatch (R-Utah) called it "the second American revolution." The federal government, Hatch said, was "waging war on the west," cheered on by "toadstool worshippers and land embalmers." Arizona Governor Bruce Babbitt perceived the rebellion as "a land grab in disguise," but most western politicians seemed to agree with Hatch, as did a newly elected Ronald Reagan. Within days after his election Reagan sent a telegram to the Nevada bill's sponsor, State Senator Deane Rhodes, averring that he too was a "Sagebrush Rebel." Shortly after the inauguration he invited rebel leaders to join him and Ed Meese and Paul Laxalt in the oval office. Once more he declared himself a "rebel" and encouraged his guests to persevere. He also promised them the cooperation of all relevant federal agencies.

The rest of the rebels rode into Washington with James Watt, Anne Gorsuch, and her husband Robert Burford, a rancher and mining engineer who ran the Bureau of Land Management under Watt. Gorsuch and Burford had been members of "the crazies," a small cadre of Colorado state legislators who fought persistently for state hegemony over federal land. Watt, Gorsuch, and Burford had all been recommended to Reagan by one of his largest financial backers, Colorado brewer Joe Coors, a member of the "kitchen cabinet." Coors had major mining interests in the West and was founder and benefactor of the Mountain States Legal Foundation, a counter-environmental law firm run by Watt.

Before he was forced to resign in 1983 after he uttered an embarrassing sequence of racial and ethnic slurs, Watt either transferred or sold over 20 million acres of federal land to states where ranchers and miners had greater influence than the U.S. Interior Department did. The eventual aim was to sell the land to private owners. Direct sale of federal land to private interests is prohibited by most of the legislation creating the federal agencies that manage it. Watt and Burford became heroes of the Sagebrush Rebels by removing land managers from the range and turning responsibility for stewardship over to ranchers, on the assumption that cattlemen would be better managers of public land than the government.

"In the end, some of the Sagebrush Rebels themselves came to question the wisdom of the movement," according to Donald Snow of the Northern Lights Institute in Missoula, Montana. "Why should the commodity interests—ranchers, loggers et al.—want to own federal lands that already offered such a bounty of subsidies? With absurdly cheap grazing fees, free minerals, and a federal road building program that gave loggers subsidized access to timber that the market would never pay for, the federal lands turned out to be a bargain too good to be true."[7]

In the meantime, the Sagebrush Rebellion stimulated a call to arms from the Wilderness Society, the Sierra Club, and other traditional conservationist organizations that had vested their reputations in fighting for public land. But the general response of the mainstream movement was to focus primarily on the cabal in the White House. Their activism, to the extent it was evident, was aimed at removing Watt and his associates from office, trusting—one supposes—that they would be replaced by significantly better people. They apparently did not consider organizing the small ranchers and landholders who were the ultimate victims of the land grab and who would eventually be mobilized into a much larger rebellion against both the federal government and the environmental mainstream.

The Wise Use Movement

Why has it become such a fetish to save energy? It's energy that built this country.

Dixy Lee Ray, Former Governor of Washington

The Nature Conservancy is a capitalist institution designed to promote socialism.

Grant Gerber, Wise Use Movement Activist

"The environmental movement has become the perfect bogeyman," according to Alan M. Gottlieb, a conservative fundraiser from Bellevue, Washington, who makes his living targeting other liberal bogeymen like Ted Kennedy and gun control advocates. Gottlieb's office is the headquarters of some of his most successful "clients," including the Center for the Defense of Free Enterprise (CDFE), the Citizen's Committee for the Right to Keep and Bear Arms, and the Second Amendment Foundation. Nothing works better in the mail, he asserts, than "an evil empire." With that in mind Sierra Club renegade and James Watt's authorized biographer Ron Arnold offered Gottlieb an ideal fundraising opportunity—counter-environmentalism. Together the two men have led the most extensive and, for environmentalists, the most disconcerting backlash to date against the movement.

Arnold's aim, in his own words, is "to destroy the environmental movement once and for all!" Arnold quit the Sierra Club 10 years ago and joined forces with Gottlieb to form the most vocal counter-conservation coalition in American history. At a Multiple Use Strategy Conference sponsored by the CDFE and held in Reno, Nevada, in August of 1988, Arnold and Gottlieb launched what is known today as the Wise Use Movement. Gottlieb raises money. And he has raised a lot of it. "I've never seen anything pay out as quickly as this whole Wise Use thing has done," he says.[8]

Meanwhile the articulate Ron Arnold serves as executive director of CDFE, where, he says, he has synthesized the tactics of Lenin with the social psychology of Abraham Maslow in formulating an ideology for a grassroots insurrection he believes will save free enterprise capitalism from the scourge of environmentalism. "The only way to defeat a social movement is with another social movement," is the lesson Arnold claims to have learned from his eclectic readings. "To call Wise Use a 'movement' may be a bit grandiose," comments investigator David Helvarg, who spent three years observing Arnold, Gottlieb, and their followers. "It is a backlash, albeit a large and well-organized backlash."[9]

The insurrection will arise, Arnold predicts, from a coalition of natural resources industry leaders, recreational-vehicle clubs, right wing ideologues, and thousands of populist, small-town, and rural citizens' groups. In his introduction to the Wise Use's bible, *The Wise Use Agenda,* Arnold calls the movement "the new environmentalism" and claims to want "productive harmony" with traditional environmentalists.

The rhetoric at Wise Use meetings and conferences however is much more confrontational. Arnold himself has publicly declared "holy war against the new pagans who worship trees and sacrifice people." Environmentalists are labeled "watermelons—green on the outside, red on the inside," and Wise Use leaders frequently complain that it is impossible to negotiate or compromise with "nature fascists" whose hidden agenda, they say, is "to destroy industrial civilization." Arnold's unhidden agenda is "to destroy environmentalists by taking their money and their members." He clearly sees himself and his "movement" as a healthy antibody created by the American body politic to destroy the virus of environmentalism.

Arnold, a top secret designer for the Boeing Corporation who worked on the Supersonic Transport (SST) project, says he broke ranks with the Sierra Club one day when a deck of logs broke loose from a Weyerhaeuser timber pile and fell into a creek. Sierra Club activist Brock Evans (now chief lobbyist for the National Audubon Society), who happened to be in the area at the time, showed Arnold some pictures of the damage. According to Arnold, he (Arnold) suggested that they call Weyerhaeuser and report the problem. But Evans allegedly said they should call the press instead and get some quick publicity for the environmental movement, which, he told Arnold, "only had two or three years of public popularity left . . . Why give that company a chance if we can smear them now?" Arnold says he was so horrified by Evans' response, that he quit the club and renounced his environmentalist leanings. Arnold has repeated the story of his conversion to many audiences. Evans calls it "horse shit of the highest order." [10]

Backed by large donations from extractive corporations such as Exxon, Louisiana Pacific, and Boise Cascade and by small donations from loggers and mineworkers, the Wise Use movement has replaced Ronald Reagan, James Watt, and the Sagebrush Rebellion as the attack force of anti-environmentalism and the mainstream environmental movement's prime bogeyman. Although some historians believe Wise Use rose from the ashes of the Sagebrush Rebellion, it is in fact as much a reaction to Sagebrush ideology as an aftereffect. Sagebrush rebels wanted to sell federal land to rich land barons; that hardly set well with the small ranchers, landless westerners, and workers in the industries that had benefited from what amounted to subsidized mining claims and virtually free grazing land at federal expense. Here was a constituency for the taking, and Wise Use organizers went after it.

The Wise Use movement has an appealing populist message for the small landowner who is told over and over by itinerant organizers that the federal government is out to steal his land and that environmentalists are conspiring in the theft. One response to the perceived federal theft is the "county movement," a Wise Use initiative calling for all control of public land to be shifted from state and federal to the county level.

Although there were traces of cultural preservation in the Sagebrush Rebellion and a lot of populist appeal, it was essentially a land grab by the rich. Wise Use, on the other hand is much broader and far more complicated. Here is how investigator Helvarg describes it:

> Scattered in hundreds of small rural towns and suburbs across America, made up of ad hoc groups in need of constant resuscitation by paid professional organizers and right wing legal foundations, kept in communication by fax nets, fliers, and conferences whose attendance never exceeds the low three figures, and prompted by friends in industry, government and the media, Wise Use/Property Rights may be neither "the most powerful organization this country has ever seen"—as reported in the Alliance For America newsletter (circulation 2,500)—nor the "brutally destructive anti-environmental onslaught" portrayed in a Sierra Club fundraising mailer. Rather it is a new and militant force on the political right that has the power to impede and occasionally sidetrack attempts at environmental protection, intimidate politicians and local activists, and polarize or misdirect needed discussions over jobs, health, and natural resources."[11]

Wise Use movement leaders, eager to assert their claims as a grassroots uprising, profess to have an active force of 5 million backed by 120 million sympathizers, an outrageous exaggeration given every known environmental poll taken in America. (See Chapter 7 for more on polls.) In fact, most of the financial support for Wise Use groups comes from large corporations in the resource extraction industries. Companies like Champion Paper and MCI fund foundations with names like the Evergreen Foundation, which in turn support groups with names like the Environmental Conservation Organization and the National Wetlands Coalition (which assists developers seeking permission to drain wetlands for commercial real estate). For example, although the Western States Public Lands Coalition, a pro-mining anti-environmental group with a populist subsidiary called People for the West, has some grassroots financial support, over 95 percent of its 1990 budget was covered by corporate donations. Moreover, all but one of the coalition's 13 board members in 1993 were mining executives.[12]

Rhetorical and ideological support for the Wise Use backlash is provided through the media by men like reactionary populist Rush Limbaugh and conservative columnist George Will, both of whom spend less time defending the tenets of Wise Use ideology than they do trashing environmentalism. "Environmentalists care so much that caring becomes a crutch that makes them feel special and more noble than the rest of us," asserts Limbaugh, who rails almost daily about "enviro-religious fanatics and eco-Nazis" on over 600 radio stations nationwide. "With the collapse of Marxism, environmentalism has become the new refuge of socialist thinking," says Limbaugh. "What better way to control someone's property than to subordinate private property rights to environmental concerns."[13]

Ron Arnold believes that the environmental assault on private property has affected the economy, reducing the gross domestic product by 6 percent annually. He also asserts that environmentalism has created many of our most pressing social problems. "When you come into a town and decide a factory has to close because it violates environmental regulations, you see behavior patterns change. Domestic violence, child abuse, the use of drugs and alcohol all go up to a new plateau as the community disintegrates. Is it the environmentalists fault? You're damn right it is."[14]

Whipped up by such rhetoric, some supporters of the Wise Use movement may become potentially dangerous. The movement has clearly attracted people from the darker side of populism, the petty thugs and weapon enthusiasts of the ultra right. A few have physically attacked activists across the country, torched their houses, and left life-threatening messages on answering machines and in mail boxes. [15] Their actions reflect a worldwide trend toward anti-environmental violence that includes physical attacks on environmental activists in Malaysia, Kenya, India, and Canada, as well as the assassination in Brazil of Chico Mendes, a rubber tapper and rainforest advocate. In Arizona the murder of Leroy Jackson, a Navajo anti-logging activist, remains unsolved. Many of his compatriots believe that one of the many anonymous people who threatened his life simply followed up on the threat.

Arnold denies contributing to Wise Use violence. "When I say we have to pick up a sword and shield and kill the bastards, I mean politically not physically." William Holmes, a retired timber baron, also denied fostering violence after he exhorted a California logging conference to initiate a "Hate Them" campaign against environmentalists, "with the single objective of destroying them as a political force." It should be a "fun program," Holmes added. "You should be prepared to kick someone in the crotch." Funding for the campaign, he cautioned, should be carefully "disguised" so that it could not be traced to the timber industry.[16]

The Sahara Club, an off-road vehicle enthusiasts group, is particularly incensed with wilderness preservationists who seek to deny motorized access to deserts, wildlife refuges, and national forests. The club maintains a computer bulletin board listing the names, addresses, telephone numbers, and license registrations of environmental activists, offering them to members with this message: "Now you know who they are. Just do the right thing; and let your conscience be your guide."[17]

Former Interior Secretary James Watt joined the philosophical jihad against environmentalism in June of 1990 by musing aloud to a gathering of cattlemen that "if the troubles from environmentalists cannot be solved in the jury box or at the ballot box, perhaps the cartridge box should be used." Given that rhetoric, there have been fewer incidents of direct violence against environmentalists in the United States than one might expect, although threats and vigilante actions are on the increase, particularly in the West.[18]

The Takings Backlash

Every man holds his property subject to the general right of the community to regulate its use to whatever degree the public welfare may require it.
Theodore Roosevelt

Takings has become the visible symbol of Wise Use.
Bill Klinefelter, AFL-CIO

In 1988 President Ronald Reagan signed an executive order mandating federal agencies to review all regulations for their effect on private property.[19] The order encouraged the oldest and most predictable adversaries in the environmental counterrevolution, the owners of private land. It was based on the legal theory that government regulations protecting land, air, water, and endangered species are tantamount to land condemnation and that owners of private land affected by such regulation should be compensated, just as they would be if the land were purchased through eminent domain. The premise is that many such regulations constitute "takings" as defined by the Fifth Amendment, which stipulates that government may not "take" private land for public use without "just compensation."

In the 1980s "Takings" became the mantra of free market environmentalists and the war cry of landholders large and small who claimed, mostly in the small and relatively obscure United States Court of Federal Claims, that environmental rules and regulations, by diminishing the development potential of their land, had affected its value. The takings initiative is backed by extractive industry associations like the American Mining Congress and the American Petroleum Institute, in concert with the American Farm Bureau and the National Association of Realtors. The main conduit of their support is Defenders of Property Rights, founded in 1991 by Roger and Nancie Marzulla, both former Justice Department attorneys in the Reagan Administration. The Marzullas, who were the architects of the Reagan executive order, became the heroes of the Wise Use movement. They work closely with the Washington Legal Foundation, the Competitive Enterprise Institute, and the "Defenders Network," a nationwide system of almost two dozen private law firms that offer *pro bono* legal support to grassroots Wise Use organizations. The network's annual meeting is sponsored by the Heritage Foundation.[20]

With 16 conservative judges sitting on the Claims Court bench—all appointed by Reagan and Bush—corporate and personal property owners began to win cases; most have been upheld on appeal. In one landmark case, *Loveladies Harbor* v. *United States,* the court found for a New Jersey developer who argued that denial of a permit to fill a wetland and build luxury houses was a taking. Loveladies was awarded $2.7 million. *Nollan* v. *California,* a frequently cited takings case, went all the way to the Supreme Court, where the majority ruled that the California Coastal Commission had no right to demand that Mr. Nollan give up a portion of his beachfront property for public access in return for a building permit. Takings again reached the Supreme Court in *Lucas* v. *South Carolina,* and again the court ruled for the plaintiff. The latter claimed he had lost the value for land (much of it underwater for most of the year) when he was denied a building permit because of public safety considerations. In his second written opinion for the five-to-four majority, Antonin Scalia remanded the case to the state for settlement and issued a footnoted invitation to bring more takings cases to the Supreme Court.

Lucas and *Nollan* were important anti-environmental victories. But they were both won by real estate developers, for whom it is difficult to engender sympathy. What the Wise Use movement needed was a little guy, which they found in Mr. and Mrs. Dolan of Tigard, Oregon. The Dolans run a mom-and-pop plumbing supply store. When they applied for a permit to expand their business, town officials agreed to issue it in return for deeding a portion of their land to the town for flood control and a bicycle path. Dolan sued the town and lost in Oregon courts. Seeing an opportunity to interpret the Fifth Amendment in favor of property owners, Wise Use lawyers took the case to the Supreme Court, which, in 1994 ruled 5 to 4 that Tigard had not demonstrated a clear connection between the public purposes and the conditions it had sought to impose on the Dolans.

Writing for the majority, Chief Justice Rehnquist sent the case back to the town for additional clarification and wrote: "We see no reason why the takings clause of the Fifth Amendment, as much a part of the Bill of Rights as the Fourth Amendment, should be relegated to the status of poor relation." Whether or not the *Dolan* decision is a clear victory for private property rights is debated by lawyers on both sides of the issue. Few doubt, however, that Rehnquist's encouragement will prompt additional litigants to bring takings cases before the high court. And no one doubts that the court has sent lower courts a clear message that takings is a legitimate claim.

At first property rights cases were a minor vexation. By the early 1990s takings litigation had become a serious problem for the government and the environmental movement. Under the takings concept, landowners began challenging the government's authority to regulate grazing and water rights, order the cleanup of toxic wastes, buy land for national parks, protect wetlands on private land, and restrict mining in wilderness areas. An implicit argument of takings litigation is that landowners have a "property right" to damage the environment as they please and that the public must reimburse them if that right is taken from them. "Oddly" says University of Montana economist Thomas Michael Power, "free marketeers somehow always assign property rights to those doing the polluting rather than those being damaged by the pollution."[21]

Even the smallest property owners have been able to take advantage of the takings craze. Imaginative lawyers with nothing better to do have convinced friends and clients that they can make money simply by buying land, applying for a permit they know will be denied on environmental grounds, and taking their case to the United States Court of Claims, where presiding Judge Loren Smith has awarded hundreds of millions of dollars to landowners who claim that environmental regulations have decreased the value of their property. In 1991, in the largest judgment to date, the federal government was ordered to pay a Wyoming coal company $150 million after the Department of Interior enjoined it from strip mining in a restricted area. The court's decision was upheld by the Appellate Court, which is hearing appeals on takings cases involving land purchased as far back as 1982.[22]

Encouraged by these victories, conservative congressional representatives have attempted to codify Reagan's executive order, which could be eliminated by a stroke of President Clinton's pen. They have introduced several compensatory "no net loss of private property" bills—legislation that would undermine not only local zoning laws but also environmental and public health legislation that took decades to win. Most of the bills would provide automatic compensation for all private property owners denied the right to improve or develop their property.

In 1993 Senator Robert Dole offered an amendment to a bill to elevate the EPA to full cabinet status. The Dole amendment would have made law another Reagan executive order, which gave the U.S. attorney general the power to halt implementation of any regulation he or she believed would result in a taking. Although the Senate stifled the Dole amendment and Congress has rejected all other takings bills so far, renewed attempts are very likely in the Republican-controlled Congress and in state legislatures.

To date Utah, Arizona, and Delaware have passed takings bills; and similar legislation has been introduced in more than half the remaining states. The Utah bill, the nation's first, was pushed through the state legislature by a vocal minority calling itself "the Cowboy Caucus." Mark Pollot, who drafted the Reagan order and is now in private practice, is promoting an even stronger piece of model state legislation that would automatically compensate private landowners who lose 50 percent or more of their land's value due to an environmental regulation, or even to denial to fill a wetland. Pollot's model bill has been introduced in 10 states, and has national environmental organizations heavily vested in Washington scrambling to retarget state legislatures, where takings bills have been passing with relative ease.

Takings is a grassroots backlash that could tie up the environmental bar for years to come, distracting talented lawyers from toxics litigation and even weightier issues. It also further illuminates the limits of legislation. It seems there isn't a loophole that a creative lawyer can't find in a bill that threatens private property or the rights of owners—even when the bill is the Bill of Rights.

Mainstream Response

The Wise Use movement is expanding while the environmental movement is shrinking. This is particularly frustrating for organizations like the Sierra Club that have spent thousands of hours and dollars investigating the anti-environment movement. The club has learned in the process that Wise Use is a front for extractive corporations, that it has built unsavory alliances with the ultra right, and that much of its appeal is based on lies. It must be especially disheartening to discover that many of the companies with whom the mainstreams have communicated in good faith over issues of land stewardship and sustainable forestry are supporting the Wise Use agenda.

The mainstream response to populist anti-environmentalism has been inappropriate. Environmentalists openly treat Wise Use supporters as if they were a bunch of rubes who don't know any better than to believe an uneducated jackass like Rush Limbaugh. The dismissive arrogance and the general elitism displayed by many environmentalists in rural areas of the West has been one of the best organizing tools of the Wise Use movement. Ironically, the same attitudes that have alienated mainstream enviros from their own grassroots are helping create Wise Use's grassroots.

Its activists are plowing the fields that environmentalists could be plowing. They are organizing in the mostly rural communities while the mainstream environmental movement continues to rely for its support on the mailing lists of liberal magazines, suburban charities, and other environmental organizations. Wise Use activists too use direct mail to raise money and build lists, but they don't confuse the process with grassroots organizing, which requires face-to-face communication in living rooms and town hall meetings. Ron Arnold and Chuck Cushman, a Wise Use orator who spends much of his life addressing small town meetings, understand that cadres recruited by direct mail are passive and their loyalty to the cause is ephemeral.

However deceptive it may be, the appeal of the Wise Use message is very real to hard-working farmers, loggers, and miners—the very people environmentalism needs to reach if it is to survive as a relevant and effective movement, particularly in the West. As outrageous and potentially dangerous as its adherents may seem at times, there is some merit in the Wise Use message, and it could be acknowledged by environmental leaders without hurting their cause. Ordinary people in the countryside are being ignored and disrespected by the same government agencies that the environmental movement has been fighting for decades.

It has never been easy to sell the environmental message to working people, particularly those who, like mineworkers, lumbermen, millworkers, and cowboys, may live near the edge of destitution. Most of these people love the land around them as much as any Sierra Club hiker does. If environmentalists worked in their communities, as Wise Use activists do, they could convince them that the property values of small farmers and landowners are actually threatened by the major supporters of Wise Use activity—the large mining, ranching, and timber corporations headquartered in Denver, Houston, and New York.

Chuck Willer, director of the Coast Range Association in western Oregon, is one of the few forest activists in the country who targets logging abuse on private as well as public land. In the course of doing so he is frequently confronted by Wise Use adherents in the small woodlot towns and villages along the Pacific coast. But he talks to them. "What the Rush Limbaugh crowd has to be told," Willer believes, "is that the owners of these forests are in jets at 40,000 feet, and they don't give a shit about you."[23] Josh Reichert at the Pew Charitable Trust, headquartered in Philadelphia, seems to understand the same principle. "American environmentalism has never been a grassroots movement." he says. "Environmental organizations are going to have to become adept at mobilizing ordinary people."[24]

Andy Kerr of the Oregon Natural Resources Council hears the message too. In 1994 he moved his family from Portland to Joseph, Oregon, a small logging town where he and other enviros were blamed for a recent mill closing. Kerr, who could paper most of the rooms in his new house with the death threats he has received, intends to "knock on doors and get involved with local issues, showing people that their concerns are our concerns."[25]

5

The Third Wave

The lion will lie down with the lamb, and toxic polluters will drink herb tea with environmental activists.

Thomas Michael Power

Environmentalists have found themselves in the position of knowing how bad things are but only capable of making a deal.

Peter Berg, Planet Drum

The third wave is the future.

Fred Krupp, Environmental Defense Fund

At a July 30, 1994, meeting of the Coalition for Clean Air in Los Angeles panelist Gail Ruderman-Feuer, a lawyer for the Natural Resources Defense Council, described herself this way: "Environmentalists used to be seen as people who try to get the toughest possible regulation legislated. Now we are sitting down a lot more with business. We are saying 'let's be more flexible' and we are looking at market approaches. Now that we are friends of business, they have become faces, not just names on pieces of paper." It was Saturday morning, but Ruderman-Feuer was dressed as if she had a court appearance.

On the same panel, a few chairs from her and dressed for a walk on the beach, sat Bob Wyman, partner in the Los Angeles law firm of Latham Watkins, which, in Wyman's own words, "represents some of the largest polluters in the L.A. basin." Wyman smiled during Ruderman-Feuer's introduction. Clearly she was saying what he wanted to hear. After Wyman introduced himself, he made a plea to reward ARCO and his other clients "for doing the right thing. They need incentives." With two such advocates, free market, "third wave" environmentalism was well represented.

Some historians of American environmentalism have recently conceived the movement in terms of three waves. The first wave, which came in the early twentieth century during the presidency of Theodore Roosevelt, ushered in the era of land and wildlife conservation. The second, sparked by Rachel Carson's *Silent Spring,* was marked by a decade of landmark environmental legislation banning or limiting the pollution of land, air, and water. The third wave broke on America's shores during the Reagan years and is still washing over us.

Courtroom to Board Room

Environmentalists have to be architects, not just complainers.
Fred Krupp, Environmental Defense Fund

If it had been the purpose of human activity on earth to bring the planet to the edge of ruin, no more efficient mechanism could have been invented than the market economy.
Jeremy Seabrook, The Myth of the Market: Promises and Illusions.

The current buzzwords of third-wave environmentalism are *market-based incentive, demand side management, technological optimism, non-adversarial dialogue,* and *regulatory flexibility*. The last phrase is lifted from a Reagan communiqué to Congress on the state of the environment: "We have expanded innovative programs which allow industry the regulatory flexibility and economic incentives to clean up pollution." *Constructive engagement,* a term borrowed from Cold War foreign affairs, has also been heard in third-wave circles.

The essence of third-wave environmentalism is the shift of the battle for the environment from the courtroom to the board room. Many of the same organizations that were once eager to take environmental offenders to court now wish to sit down and hammer out a deal that allows each party to declare victory and appear green. It is an attempt to "get rid of the combativeness between activists and corporations," according to Don Walukas, director of Environmentally Conscious Manufacturing for the Center for Manufacturing Science, a corporate-supported think tank. In fact, third-wave environmentalism represents nothing so much as the institutionalization of compromise.

Since the early 1980s advocates of the third wave have risen to the top of some national environmental organizations (for example, Jay Hair at the National Wildlife Federation and Fred Krupp at the Environmental Defense Fund). Others—like William Reilly in the Bush administration—received powerful government appointments. The mainstream movement is not, however, of one mind about third-wave environmentalism, and the internecine debate is heated. The outcome will shape American environmentalism for the next decade or two. If the third wave prevails, it could destroy forever any hope of meeting the environmental objectives laid out in the preamble to the National Environmental Policy Act of 1970. It is not, however, likely to become the dominant paradigm. It will probably remain an experiment that produces a few worthwhile results but is no more than a minor sideshow to twenty-first-century environmentalism.

Proponents of the third wave like former senators John Heinz and Timothy Wirth, environmental leaders Krupp and Hair, and most corporate executives believe that government should use "market-based incentives" instead of regulation to induce industry to pollute less. What exactly is meant by "market-based incentive," of course, remains as vague and undefined as Adam Smith's invisible hand.

In what sounds like a quotation from the Heritage Foundation's *Mandate for Leadership,* Jay Hair takes a stab at defining third-wave environmentalism. "Our arguments," he asserts, "must translate into profits, earnings, productivity, and economic incentives for industry." He firmly believes that "reaching out to the business world [and] enlisting the entrepreneurial zeal, the proven expertise, and the enlightened self-interest of America's private sector" will result in significant environmental progress.[1] A report cosponsored by Senators Heinz and Wirth adds more specifically that "utilizing market forces and economic common sense to achieve environmental goals entails removing market barriers and government subsidies which promote economically inefficient and environmentally unsound practices."[2]

The EDF's Fred Krupp assumes that Americans do not want to choose "between improving our economic well-being and preserving our health and natural resources." He calls for a strategy that "harnesses the profit motive and introduces new incentives to get business to do the right thing in the first place." Krupp pointed to market-based incentives as "an area where the environmental movement and the Bush Administration can find common ground."[3]

Such sentiments were understandably popular at the Reagan-Bush White House and with members of the Task Force on Regulatory Relief (chaired for a time by Vice President Bush) and its later manifestation, the Task Force on Competition (chaired by Vice President Dan Quayle). William Reilly, former president of the World Wildlife Fund and Bush's EPA administrator, was also pleased. "The time has come," he wrote in a 1989 article entitled "The Greening of EPA," "to consider applying market incentives/pollution prevention approaches to environmental programs across the board."[4]

The faith that Reilly, Krupp, Hair, and others on the environmental right share is that the privatization of cost-benefit analysis would have salutary environmental results. It is a notion that, naturally, seemed like economic common sense to the heaviest polluters—many of whom began to support conservative environmental organizations with generous donations in the late 1980s.

The logic of market-based incentives is rooted in several assumptions about the market's relationship to the environment. The most important is the notion that production decisions should remain in the private sector and that removing market barriers and government subsidies that promote environmentally unsound practices will allow the mechanisms of the marketplace to motivate industries to make environmental protection profitable. Another implicit tenet of the third-wave ideology is that all non-fraudulent businesses and industries deserve to exist, even if their technologies or products are irreversibly degrading to the environment (or as long as efforts are made, as required by law, to limit the degradation). Finally, it holds that only environmental scientists and lawyers possess the necessary expertise and sincerity to negotiate environmental issues fairly with corporations or government agencies. The general public and its representatives outside those professions are simply too ignorant and self-interested.

Third-wave optimists maintain that allowing manufacturers to reduce emissions and waste disposals gradually while new, less-polluting technologies are developed will eventually produce a cleaner industrial environment. The unfortunate consequence of this reasoning is that while it is being proven valid or invalid pollution is continuing. The fact that the rate per individual polluter may decrease creates a misleading impression of progress.

Moreover, such schemes may founder if they do not take levels of production into account. For example, DuPont, Monsanto, and Dow might each receive an economic incentive to reduce their emissions by 50 percent over 10 years—a figure far above any demanded by the mainstream environmental organizations. Even if they do so and industrywide production of the same chemicals doubles during the period, nothing will be gained. To be effective, therefore, incentives must outpace the growth of polluting industries and arbitrary ceilings for all polluters must be established. The problem here is that the system is only as strong as the ceilings, and there are so many exceptions, exemptions, and loopholes in the optioned incentives that little actual progress is made. The results—an increasingly polluted environment—are evident.

A Market for "Bads"

Adam Smith's invisible hand can have a green thumb.
President Bill Clinton

Third-wave environmentalists, particularly Fred Krupp and Dan Dudek at EDF, take great pride in their invention of *emission reduction credits* (ERCs). They are also known as *pollution allowances, pollution permits, pollution credits,* and—to less accommodating air quality activists—*pollution rights* or *cancer bonds*. The innovation was introduced in an amendment to the 1990 Clean Air Act by Dudek "after dozens of meetings with White House and Environmental Protection Agency staff."[5] It is based on a concept called "trading under a cap." A national limit on a toxic emission, sulfur dioxide for example, is set by the EPA, and the total is divided and allocated to polluters. Utilities and other emitters of SO_2 can then trade emissions among themselves as emission reduction credits.

The buying, selling, and banking of ERCs was a blessing for serious air polluters frustrated by the costly regulatory standards of state and federal governments. A federal requirement to halve 1980 levels of SO_2 emissions by the year 2000 was especially worrisome. (Sulfur dioxide is the most critical emission, not only because of its sheer volume but because it causes acid rain and compromises the ability of the human respiratory system to withstand other pollutants.) By allowing companies to "bank" the allowable smog-forming vapors they are not pumping into the atmosphere and sell them to companies unable to meet emission standards, the government hopes to lower toxic emissions nationwide without driving undercapitalized public utilities out of business. Whether it will actually result in lower emissions is anyone's guess. Melanie Griffin at the Sierra Club was willing to take a chance. "If the cap works it will be great," she said as the provision was being considered for the Clean Air Act of 1990. "If it's going to get a bill through the House and Senate then we might as well give it a try."[6]

Pollution credits were institutionalized as a compromise provision of the Clean Air Act of 1990. Credits for sulfur, nitrogen, carbon dioxide, and employee trip reduction (ETR) are now considered as securities or commodities. They are even traded on the Chicago Board of Trade, which recently won approval from EPA to conduct an annual auction of pollution credits.[7] Ultimately, the board hopes to offer pollution futures, which anticipate future issuance of emissions credits. Among the likely buyers are cities that have lost defense contractors and hope to attract new, potentially polluting industry by offering low-cost pollution credits well into the future.[8]

Pollution credits are designed to reduce air pollution gradually, reward companies for diminishing emissions, allow undercapitalized utilities to stay in business, and spread the costs of emission a bit. They make some sense in theory, but as an environmental protection method they are flawed. Because allotment of ERCs is based on previous fuel use and past emission rates companies that have been the most wasteful receive the largest allowances. Further, inclusion of provisions allowing companies to base their allotment on future emissions diminishes the motivation to implement pollution-free technologies and sends the country's worst polluters a comforting message. Creating a marketable—and profitable—property right to emit pollutants lessens incentives to cut or lower emission ceilings. Even in a best-case scenario, attainment of cleaner air is delayed even further.

In effect, Congress has created a valuable financial commodity out of the right to emit toxic gases and, instead of selling it, donated it to the electric utility industry. The public, therefore, receives no assured benefit from the program, not even the mere dent in the budget deficit such revenues could provide.[9] The free handout has prompted other industries to push for extension of the pollution credit system to certain volatile organic chemicals as well as to other air and water pollutants.

In the acid rain program the practice has backfired badly. Pollution credits sold by eastern utilities to their Midwestern counterparts, which needed them in order to burn high-sulfur coal, eventually created vast sulfur dioxide plumes that were carried back east on prevailing winds. As a consequence, even more acid rain has fallen on fragile northeastern lakes and forests in the United States and Canada.[10] The practice has also led to the creation of what are known as pollution "hot spots." ERCs sold by a clean, well-financed utility to a company hundreds of miles away that is unable to meet its emission standards create an area of toxic and greenhouse gases down wind from the dirty plant. All too often the new hot spot is a poor and less politically influential region of the country.

An opportunity to study ERC-induced hot spots exists in the Los Angeles Regional Clean Air Incentives Market (RECLAIM) program. Launched in early 1994, RECLAIM permits ERC trading within the heavily polluted Los Angeles basin and is seriously considering trading credits for volatile organic chemicals (VOCs). The trade would create the potential for a truly toxic hot spot with direct and immediate effects on human health.[11]

From a global standpoint, the situation is no less serious. Tons of greenhouse gases—sulfur dioxide and nitrogen oxide—are still being belched into the atmosphere; only the locations of the pollution sources have changed. As long as growth of an economy based on petrochemicals continues the market approach to environmental protection can, at best, create a standstill. At worst it will be completely self-defeating of the goals it purports to support.[12]

Another problem with market incentives is that they make no provision in standard corporate accounting for environmental protection; *not* polluting, therefore, has no positive value on a balance sheet, income statement, or tax return. A firm receives no financial credit, for instance, when it chooses not to release effluents into a stream. Only the debit side of pollution control—reflected in the cost of storing or transporting toxic effluents, installing scrubbers, buying catalytic converters, and so on—is entered in the books. This fact defeats the whole private cost-benefit argument. If there is no financial benefit attached to not polluting, the only real incentive often is to pollute. Without taxes, fines, or other financial penalties to create a negative incentive, polluting becomes very profitable.

It is true that in some instances pollution prevention has been found to increase profits, and free marketers and their third-wave supporters are quick to cite such cases. Since 1975, for example, the 3M Corporation has saved over $500 million through energy conservation and avoidance of hazardous waste. Dow Chemical boasts of similar results in Michigan and Louisiana, and there are others. These cases do not, however, alter the fact that polluting pays off handsomely for other manufacturers and that some public control of industrial production is required.

It was not the marketplace, after all, that induced oil refiners to remove lead from gasoline or auto manufacturers to install the catalytic converters that lowered lead and sulfur emissions by 90 percent. It was state- and federal-mandated air quality and emission standards. Detroit has consistently resisted all such standards, even though they apply equally to all manufacturers, foreign and domestic. If left to their own devices, it is highly unlikely that the auto industry would have found economic incentives to lower exhaust emissions.

Another third-wave incongruity appears whenever a clean productive technology is developed to produce a product that itself becomes a pollutant. Invariably the new technology is promoted as an environmental advancement, and the life cycle of the product is ignored. The immediate community around a gift wrap manufacturer in Tennessee might have good reason to rejoice when solvent-based dyes are replaced with water-based inks. The company saves money by not having to store toxic wastes, and local ground water is protected from leaky tanks. However, the air around incinerators in communities where the used gift wrap is burned and the landfills where it is buried will still contain the heavy metals, dioxins, and other contaminants found in exotic gift wrap. Misleading publicity about minor technological fixes and environmental remedies of limited value like this has become so pervasive that 10 state attorneys-general have ruled that corporations may not promote such innovations in their states without full life-cycle assessments.

The Acid Rain Roundtable held in 1991 opened another window on third-wave environmentalism. The roundtable brought together officers of some of the major sulfur-dioxide- and nitrogen-oxide-producing utilities of the northeast and representatives from several mainstream environmental organizations. Offered the right to explore their own incentives, privately owned electrical utilities operating coal-burning plants in the Midwest reported they considered every other alternative to applying technology to lower emissions. Rather than placing scrubbers on their stacks, many simply bought the right to pollute from another utility. Such decisions did virtually nothing to alleviate acid rain. Accepted without challenge by environmentalists in attendance were concepts like cost-benefit analysis, graduated reduction of emissions, and a mandate to burn low-sulfur coal—a political expedient that allowed utilities to avoid installing scrubbers (thereby raising energy prices) and affected the welfare of only a few thousand relatively powerless miners of high-sulfur coal in Appalachia and the Midwest.

Coal miners and other blue-collar environmentalists who struggle against polluters close to home understand the follies of constructive engagement far better than their well-educated counterparts in Washington. They complain that third-wave environmentalism represents nothing less than a massive capitulation to polluting industries. They are particularly incensed by inventions like the aforementioned pollution credit, which assumes a preordained right to pollute their neighborhoods.

Among the episodes of mainstream insensitivity to local issues grassroots environmentalists cite is Jay Hair's now-notorious mediation between EPA Director William Reilly and Dean Buntrock, CEO of Waste Management Corporation (WMX). The mediation took place soon after the company was publicly rebuked for "endangering and degrading the environment" by the Environmental Grantmakers Association, a consortium of over 100 foundations and other philanthropists.

Hair knew that Reilly was concerned that some states were setting standards higher than federal standards and too high for the waste management industry to meet. Buntrock had told Hair that WMX was having this trouble in South Carolina. The three met over breakfast in Washington to discuss that state's waste-disposal regulations and the stalled approval of a toxic-waste incinerator Buntrock wanted to erect in a small New Jersey town. Reilly agreed to approve the incinerator and to change EPA policy to solve the company's South Carolina problem. No representatives of the affected communities were invited. To preserve his own environmental credentials, Jay Hair publicly opposed the policy change but, after a congressional inquiry, admitted his involvement in the negotiations had been injudicious.

Another problem with third-wave environmentalism is that the term *market incentive* has become a convenient, all-encompassing label for every possible concession to free enterprise. Thus the pollution tax, a fairly sensible idea, is approved by non-polluting businesses and called by some a market incentive. Others apply the term to the Toxic Release Inventory reported annually by the EPA and based on periodic corporate reports. The rationale for gathering the information is that the public will be less likely to buy the products of a company with high toxic releases. That outcome can only come about, of course, if the public is informed of the release, which is unlikely to occur except when a sudden leak requires evacuation, gets the attention of the media, or leads to direct action such as a consumer boycott—events that few business leaders would describe as acceptable market incentives. Market *dis*incentives don't count.

The standard defense offered by third-wave environmentalists, when accused of compromising with the government and sitting down with industry, is that the critics have it backwards. It is government, they say, that is compromising, and it is industry that is sitting down with environmentalists who would otherwise take them to court and delay efforts at environmental protection with time-consuming litigation. There is a little bit of truth to that. It was Dean Buntrock, after all, who invited Jay Hair to breakfast, an invitation that certainly must have given Hair a temporary sense of personal triumph. But Buntrock certainly knew that Hair was not about to take him to court over an incinerator. Buntrock, after all, sits on Hair's board of directors at NWF.[13]

The notion that government and industry are somehow on the run or are intimidated by the specter of litigation is specious. It ignores years of counter-regulatory plotting on the part of industry and underestimates the brilliance of corporate strategists. By offering to negotiate or mediate might they not simply be practicing political *aikido*—yielding to the first lunge to gather strength and balance? Are they not gaining more from constructive engagement than they might from another bill or a well-prosecuted lawsuit—particularly when they are seen to be negotiating in good faith with the representatives of a million or more "environmentalists"?

The often-expressed notion that third-wave environmentalists are "shaping the debate" is self-deceptive hype and is sometimes accompanied by the self-congratulatory observation that corporate adversaries are now calling themselves environmentalists. The truth is that industry only approaches environmentalists when they believe they can get a better deal than they might from a regulatory agency, a judge, or a legislature. On several occasions since the near victory of California Proposition 130 (Forests Forever) in 1990 representatives of the timber industry have approached the Sierra Club with offers to work together for the "good of the environment." If it had passed the forest referendum would have stopped all logging of old-growth timber and mandated statewide sustainable-yield timber harvests. Big Green (Proposition 138), a companion initiative on the same ballot, would have codified every industrial and agribusiness nightmare into a single statute and made California the most environmentally strict state in the union. Although both were defeated, the very existence of the propositions encouraged the industry to use a different business strategy—an old legal tactic of tort litigation: approach the plaintiff before the plaintiff approaches the bench. It is sound strategy, and it's aggressive—like yielding in *aikido*. With compliant third-wave enviros, it pays off.

There are other ways to soften the opposition. Not surprisingly, some of the worst environmental offenders—for example, DuPont, Chevron, Monsanto, Mobil, and WMX—have become some of the largest environmental donors. And organizations led by third-wave enthusiasts, notably NWF, the Conservation Foundation, National Audubon, and the World Wildlife Fund, are among the largest recipients of corporate largesse.[14]

Jay Hair at NWF invented a particularly creative mechanism to attract corporate encouragement of third-wave principles: the Corporate Conservation Council. DuPont, Monsanto, ARCO, Ciba Geigy, and others pay $10,000-annual membership fees for the right to attend occasional off-the-record enviro-seminars and field trips to places of special environmental interest. The council's quarterly newsletter, *Conservation Exchange,* is sent to CEO's of the Fortune 500, foundation leaders, and business schools—sources, Hair hopes, of future third-wave adherents.

Among third-wave environmental organizations, only EDF maintains a rigid screen for corporate donors that would compromise its negotiating positions or damage its reputation as an independent advocate for the environment. Fred Krupp was particularly careful not to accept offered support or coverage of expenses from McDonald's in connection with a joint study of fast-food waste streams (*see* Chapter 6).

The closer mainstream environmentalists get to corporations and regulators the more difficult it becomes to maintain their independence and identity as adversaries. There are already early signs, expressed in wardrobe, language, and office decor—particularly at EDF and the National Wildlife Federation—that some enviros are falling prey to the "Stockholm syndrome," a psychological condition in which prisoners-of-war come to embrace the culture and ideology of their captors. Third-wave environmentalists not only accept the notion that production decisions should be market driven and left to private interests, they also seem to believe that technology, like science, is value-free, objective, and should remain beyond the domain of public influence. These entrenched precepts of American corporate ideology appear to hold Jay Hair, Fred Krupp, and their colleagues captive. By embracing them, third-wave environmental leaders discourage the participation of their members and the public at large in the assessment of the environmental impacts of new technologies.

Not all environmentalists have gone along with the third wave. "If you were trying to handle drug problems in your community, you wouldn't be saying: 'Let's try to work this out with the drug dealers,'" Greenpeace Director Peter Bahouth told the *Wall Street Journal*. He is not a surprising source for such sentiments perhaps, but even the comparatively conservative Sierra Club Chairman Michael McCloskey has expressed misgivings about EDF's "mystical faith in the elegance of market-based solutions."[15]

It is true that Hair and Krupp are but two of more than a dozen environmental leaders active in the mainstream movement and that some of their colleagues are openly critical of their policies and actions. But together, or in chorus with one or two other third-wave enthusiasts, they can do a lot of damage by helping corporate polluters create the public impression that they are a lot greener than they really are. The more successful the third wave appears, the more ordinary citizens seem to believe it can solve a number of marginally significant problems. As third-wave national organizations promote their triumphs and virtues through direct mail, and as corporations do so in long advertising supplements in mass media journals (carelessly mixing in a few small victories with a lot of false assumptions about the benefits of the market), members of the public who have not yet suffered the effects of direct exposure to odious pollutants begin to believe that the problem is being remedied and become complacent. A $25 annual donation to a mainstream organization becomes their sole participation in environmental protection. A few million such donations create a handsome budget for third-wave environmental organizations that are often dependent for their survival on the generosity of passive and gullible donors.

George Bush substantiated his self-characterization as "the environmental president" by pointing out that the Clean Air Act of 1990 and the EPA's revised Pollution Prevention Policy Statement (PPPS) were hammered out in dozens of meetings among his staff, industry leaders, and representatives of the Environmental Defense Fund. Bush called the EDF's Fred Krupp, who was accompanied to the White House by at least 15 other Washington-based environmental leaders, "my kind of environmentalist." What Bush did not acknowledge, of course, was that the PPPS came out of those meetings seriously crippled and not one single Clean Air "reform" that came out with it did anything to prevent pollution. The control strategy of restricting air pollution to some arbitrarily mandated "acceptable" levels remained intact. Had Ronald Reagan realized how compliant and solicitous mainstream environmentalists could be, he might have invited more of them to tea in the Rose Garden.

Yo! Amigo!!

It is the good fortune of many to live distant from the scene of sorrow.
Tom Paine, Common Sense

Yo! Amigo!! We need that tree to protect us from the greenhouse effect.
Scott Willis, cartoonist[16]

While third-wave environmentalism was becoming established in America, American environmentalism began to go international. The globalization of environmental politics prompted many mainstream organizations to create international divisions, solicit members overseas, open offices in foreign capitals, and dispatch emissaries and consultants to the less-developed world. Overseas, as at home, American environmental leaders have sought new ways to reason and negotiate with environmental offenders.

Their intentions were as noble as those of the early missionaries, European colonists, and development advocates who preceded them. The multinational invasion of these earlier enterprises had resulted in pollution by colonial industries, rapacious mining practices, and the importation of toxic chemicals and waste from developed countries. Present-day third world countries, burdened with debt and coerced into rapid resource exploitation and commodity exporting to repay it, are on the verge of becoming global polluters in their own right. Major ecosystems are at risk. Northern environmentalists often complain of the southern hemisphere's obsessive haste to develop. Such critics are insensitive to the plight of countries like Ecuador, Nigeria, and Indonesia that are forced into "structural development programs" by Bretton Woods one-dollar-one-vote institutions like the World Bank, where the United States and its trading partners exercise enormous authority over lending and development policy. To pay back World Bank and IMF loans many third world countries must accept hastily deployed, environment-altering technologies imported from the industrialized world under the rubric of free trade. Others are coerced into rapid exploitation of export food crops, minerals, and timber to pay debt. Those countries compelled to develop commercial forest industries have experienced particularly harsh social and environmental consequences.

Northern environmentalists claiming superior understanding of ecological sciences and technology have recently drawn some curious conclusions about the role of the South in global pollution. In 1990, for example, the Washington-based World Resources Institute (WRI) issued a study purporting to show that underdeveloped nations of the global South—especially China, India, and Brazil—pumped as much carbon dioxide into the biosphere as the developed countries of the North. The assertion was challenged by the Center for Science and Environment in New Delhi and numerous credible ecologists and earth scientists in both hemispheres. By most measures, even of potential buildup of greenhouse gases, the finding seems absurd. Yet such an assumption, offered by a prestigious environmental think tank, provides the North an excuse not to shoulder a larger share of blame and responsibility for global warming. As a justification for environmental imperialism, it will surely be used to formulate aid and multinational lending policies for years to come. (WRI, whose support is almost entirely corporate, recently joined the Green Group. Rafe Pomerance, a former WRI officer, was appointed to a State Department post by Bill Clinton.)

The irony of WRI's North/South thesis is that close examination of almost any example of environmental degradation in the southern hemisphere uncovers the complicity of a northern multinational corporation. In many cases, a mainstream American environmental organization will discover the problem first and negotiate some sort of accommodation with the violating corporation. This is what happened in 1990, when the Natural Resources Defense Council mediated between Conoco and the Ecuadorian government over controversial oil-drilling rights in the Amazon rain forest. More recently WWF worked with Chevron on construction of an oil pipeline through the wetlands of Papua New Guinea.

The NRDC–Conoco negotiations illuminate the hubris and ecological imperialism of some American environmentalists.[17] After visiting Ecuadorian environmentalists and the tribal homeland of the Huaorani people, who occupy the Oriente where Conoco very much wanted to prospect for oil, NRDC International Program Director Jacob Scherr and staff attorney Robert Kennedy, Jr. wrote a letter to Edgar Woolard, chairman of DuPont (which owns Conoco). In it they condemned Conoco's plans for the Oriente. For Woolard the letter was added evidence of worldwide environmental opposition to Conoco's drilling plans. If he couldn't find some support for the exploration there, he said in late 1990, he would consider closing down the entire Ecuadorian operation. Before giving it up, however, he arranged a meeting with NRDC in New York.

At the meeting Woolard told Scherr and Kennedy that because Ecuador was an oil-dependent nation, drilling was inevitable; if Conoco didn't drill there someone else would. The young lawyers commenced private negotiations with Conoco, convinced that it was better to have Conoco do the drilling than another firm. Without consulting the Huaorani people or the appropriate Ecuadorian environmental organizations, Scherr and Kennedy struck a deal: Conoco could drill on the Huaorani reserve in return for a $10-million donation to an Ecuadorian foundation created by NRDC and Cultural Survival, an indigenous-rights group based in Cambridge, Massachusetts. Ecuador's Fundacion Rio Napo had little or no contact with the Huaorani people and had no authority to negotiate on their behalf.

La Campana Amazonia por la Vida, a representative coalition of 13 environmental and human rights organizations in the Oriente, complained bitterly that "NRDC has jeopardized two years of work by the Ecuadorian environmental and indigenous communities to fend off Conoco's oil development plans. In pursuit of their goals NRDC misrepresented the views of Ecuadorian environmental organizations [and] intentionally deceived Ecuador's indigenous people about their true aims and the extent of their dealings with Conoco." In an open letter to the American environmental community, La Campana said: "In general, we welcome the opportunity to collaborate with our North American colleagues. We greatly appreciate and fully support their efforts to lobby the U.S. Congress, the multi-lateral development banks and the multinationals on behalf of the Amazon biosphere and its inhabitants. But we are confronted with two North American organizations which have purposely misled Ecuadorians and who insist on negotiating with an oil company, without our consent, and without respecting our right to articulate our own needs and aspirations."[18]

Although NRDC and their emissaries have been roundly criticized by fellow American environmentalists (many of whom have also been to Ecuador and meddled in local affairs) and by Ecuadorian activists, the effect of this particular deal may well remain. It could create a dangerous precedent, smoothing the way for many more decades of degrading resource exploitation in the southern hemisphere. It could also spread the contagion of unofficial compromise to third world environmental organizations eager to seek the advice and counsel of "experienced" American environmental groups. On the other hand, it could serve as a constructive object lesson that motivates American environmental organizations to diversify their staff, improve North-South communications on environmental issues, and re-examine the wisdom of third-wave environmentalism.

Accepting Risk Assessment

Man has lost the capacity to foresee and forestall. He will end up destroying the earth.
Albert Schweitzer[19]

No single development of the past 10 years illustrates the bankruptcy of mainstream environmental imagination better than the tacit endorsement of quantitative risk assessment as an acceptable tool for environmental management and policy. While not a third-wave invention *per se,* risk assessment is enthusiastically accepted by most third-wave enthusiasts and is equated, in the minds of grassroots enviros, with other symptoms of the mainstream's malady.

When risk assessment first appeared, it was opposed by most environmental leaders. Determining policy and standards by calculating the risk of specific compounds or chemicals in ratios such as "parts per million," "parts per billion," or "cancer deaths per hundred thousand" was seen as morally suspect. Yet mainstream environmentalism, leader by leader and organization by organization, has all but embraced risk assessment as a legitimate measure of policy. Weighing the ill effects against the benefits of a given pesticide, it embraces the concept of "negligible risk"—the chance that fewer than one in a million consumers will develop cancer from exposure to a pesticide. Negligible risk does not, and probably cannot, calculate the cumulative effects of many one-in-a-million chemicals as they collect in the environment, reach the food supply, and eventually interact in human tissue.

According to the National Academy of Sciences pesticides are directly responsible for just over 2 percent of cancer deaths in the United States. That's almost 10,000 deaths. The complete anonymity of the 10,000 people who die every year from pesticide-induced cancers makes it politically possible to apply risk assessment. (Of course, if one person kills a spouse by spiking his or her food with pesticides, he or she would be tried for murder.) What makes the 10,000 deaths acceptable, to some even morally acceptable, is the fact that the mainstream environmental organizations to which they look for guidance are no longer outraged by the process that allows them to happen.

Granting that government cannot possibly assure a zero-risk society, and that assessing the risk of chemicals is a lamentable but unavoidable method of differentiating one carcinogen from another, the question remains: What should be the role of environmentalists in this process? Should environmental advocacy groups be party to the compromise solution or should they be fighting relentlessly for a healthier environment—even at the risk of defeat?

Grassroots anti-toxics activists believe that all environmentalists should oppose all forms of risk assessment. It is the only way, they say, that industry will ever be forced to change its production processes. It's one thing for the EPA to cast holy water on a cost-benefit calculus that weighs a certain number of human deaths against the cost of preventing them, quite another for an environmentalist to do the same thing. Nevertheless, mainstream leaders have gradually shifted toward a pragmatic position on risk assessment. Their organizations are behaving more like government agencies than public advocacy groups created to defend the environment and protect public health.

Conclusion

There is an imminent danger of third-wave environmentalism being institutionalized in America. In September 1994 *Fortune* magazine reported that three former EPA administrators: Russell Train, William

Ruckelshaus, and William Reilly were proposing "a new federal agency of scientists and economists to assess risks, weigh costs and benefits and help government determine regulatory priorities." *Fortune* suggested that another option might be for "the President to create a high-level commission and charge it with hammering out a consensus on how to interject greater rationality and cost-effectiveness into environmental rule making. Such a body's members would include Administration officials, Congressmen [*sic*], business and labor leaders, and environmental activists." [20]

Market-based incentives should not be ignored or discarded. But it should be clear to any experienced environmental advocate invited to join such a commission that market incentives will not work in the absence of socially based disincentives. Incentives and disincentives should not be confused for the sake of promoting third-wave environmentalism—in the United States or in Ecuador. A pollution tax is NOT a market based-incentive. Many environmentalists, fortunately, are aware of this. Lawyer Richard Ayres, a founder of NRDC now in private practice in Washington, still questions the moral premise of third-wave environmentalism. He was particularly piqued by the creation of pollution credits by environmental leaders: "There was no mention from environmental leaders of the fact that the 1990 Clean Air Act was giving people the right to pollute. The air is a public resource. It should not have been given away to private operatives. Congress, with the cooperation of environmentalists was," according to Ayres, "giving away a public resource. The morality of that was never discussed."[21]

The intention of third-wave environmentalism is to protect the environment while preserving economic prosperity and price stability. But the hidden costs of cheap lumber, cheap energy, and cheap gasoline are acid rain, vanishing and extinct species, loss of arable farmland, and future generations of deformed children. Not until those "externalities" are dealt with in an open and democratic way will third-wave incentives make sense. As Thomas Power and Paul Rauber point out, "a transfer of wealth of this magnitude cannot take place without considerable conflict. In the past such power shifts have required revolutions."[22] "If free marketeers based their program on charging the true environmental costs for all resources used, the environmental movement would sign up en masse," according to Power and Rauber. "But ideological free market environmentalists too often seem more concerned with the market than with the environment. They tend to feel that equity—the distribution of access to scarce resources, or the right to a clean and healthful environment—is less important than economic efficiency and property rights."[23] While the marketplace may supply some relatively peaceable and cheap remedies for many of our ecological ailments, free-market environmentalism can never be a panacea.

The worst aspect of third-wave environmentalism is that it is essentially anti-democratic. Environmental protection, to the extent that it is achieved at all, is won through negotiation among the powerful. When Fred Krupp, director of the Environmental Defense Fund, cuts a deal with General Motors over automobile emissions there is no public participation. When he enters that board room in Detroit whom does he represent? The 36 members of the EDF Board? The 120,000 passive contributors? The donor foundations? Himself, or some vague principle he believes will benefit the environment? More important than these questions is whether or not he represents the public. And if he does, where was the public hearing?

Political movements never progress continuously. They suffer delays and setbacks and even lie dormant for considerable periods—perhaps to rest, heal wounds, build strength, or regroup. Those who observe with dismay the present state of American environmentalism can find consolation in the history of movements that have reawakened revitalized and gone on to make great strides. We can hope that future historians will see the third wave as such a temporary period of dormancy for the American environmental movement. Perhaps they will judge the attempt to reconcile environmentalism with corporate capitalism as an unavoidable strategy in a conservative era that still saw corporate capitalism as something close to a religion.

6

Environmental Justice

I think what's happening is that people are taking back the power to govern. It's not just symbolic power, it's real power.

William Ruckelshaus, Chief Executive Officer, Browning-Ferris

We're all in the same sinking boat, only people of color are closest to the hole.

Deeohn Ferris, Lawyers' Committee for Civil Rights

Traditional American environmentalism never had one single, clearly stated goal, or an all-encompassing slogan like "equality under law," "peace in our time," or "justice for all." It has become, in fact, a movement of many agendas, some of them conflicting. Thousands of organizations believe that their own agenda, issue, or domain—whether wilderness preservation, managing resources, saving endangered species, cleaning up toxic pollution, limiting population, preventing ozone depletion, or ensuring clean air and water—is the all-important issue of the movement. They could not all be right, of course—unless there is a common thread running through their various interests, a theme that somehow makes them all equally important. Perhaps that thread is justice.

In a nation built on notions of equal opportunity and equal rights, the environmental imagination must include the premise that the environment belongs to us all; that we share equally its life-sustaining attributes and whatever degradation we impose upon it. Until quite recently this basic point has somehow escaped most American environmentalists. Dorceta Taylor, a professor of sociology in the School of Natural Resources at the University of Michigan points to the missing piece in the environmental puzzle:

> If it is discovered that birds have lost their nesting sites, environmentalists go to great expense to erect nesting boxes and find alternative breeding sites for them. When whales are stranded, enormous sums are spent to provide them food . . . When forests are threatened, large numbers of people are mobilized to prevent damage. But we have yet to see an environmental group champion human homelessness or joblessness as issues on which they will spend vast resources. It is a strange paradox that a movement that exhorts the harmonious coexistence of people and nature, and worries about the continued survival of nature (particularly loss of habitat problems) somehow forgets about the survival of humans, *especially* those who have lost their habitats or food sources. If this trend continues a vital piece of the web of survival will be missing.[1]

If the overriding objective of environmental activism is protection of the health of the entire environment, the traditional environmental movement was no more than half a movement. Limited from the start, it was almost obsessively oriented toward wilderness, public land, and natural-resource conservation. Only recently have American environmentalists paid much attention to the immediate environment of most humans in society, or considered the air inside a small furniture factory in south central Los Angeles as environmentally significant as the air in the Grand Canyon. The simple shift in emphasis from the natural to the urban domain has transformed American environmentalism into something very different from its traditional form. It could save the movement.

The Rise of Environmental Ad Hocracy

Men took on the state and left the care of civil society to women.
Manuel Castells, The City and the Grassroots

Our goal is to plug the toilet.
Lois Gibbs, Citizens' Clearing House for Hazardous Wastes

As the mainstream, conservation-oriented movement grew and prospered, a parallel movement was developing in the United States. At first, operating as an adjunct to the labor, public health, and civil rights movements, it was not seen as an *environmental* initiative. The two environmental movements could not be more different, or separate; saying that they are as different as black and white is truer than it sounds. The mainstream movement has always been about as white as any social movement in the country's history, whereas the grassroots anti-toxics activism that rose in its shadow succeeded, where the mainstream failed, in drawing into the environmental struggle people of every color. The two movements also differed in gender—mainstream leadership being predominantly male, the grassroots predominantly female. That too is changing, in both movements.

The central concern of the new movement is human health. Its adherents consider wilderness preservation and environmental aesthetics worthy but overemphasized values. They are often derided by anti-toxic activists as bourgeois obsessions. People are drawn to grassroots environmental politics because they fear for their lives and those of their children. Although they tend to be patriotic and conservative in most matters, they are generally suspicious of electoral politics, cynical about government, and wary of large organizations, no matter how civic minded and well meaning they may appear. Most would never have called themselves environmentalists before they became active, and some still refuse to do so. Their motivation is protection of family and neighborhood, although many develop an appreciation of traditional environmental goals—and even a global perspective—after joining the fray and meeting experts from the Natural Resources Defense Council, Greenpeace, and others who come to their assistance.

In 1978 children in a blue-collar housing development in Buffalo, New York, called Love Canal started telling their mothers that their feet burned when they played barefoot on the lawn. At about the same time a black ooze began seeping through basement walls throughout the community. Citizens complained to the city of Buffalo about the ooze, skin irritations, and the fact that rocks dug up in the neighborhood exploded when dropped on a hard surface. Some mentioned that dogs and cats had been losing their fur. City officials, who knew the suburb had been built on a landfill operated by Hooker Chemical Company, ignored the complaints.

About the same time Michael, the three-year-old son of Love Canal resident Lois Marie Gibbs, developed a serious respiratory problem. Alarmed by her son's affliction and the unusual medical problems plaguing other children on her block, Gibbs organized the Love Canal Homeowners Association and took their complaints to the state capital in Albany. When state epidemiologists visited Love Canal they discovered abnormally high rates of birth defects, miscarriage, epilepsy, liver abnormality, rectal bleeding, and headaches.

In August of 1978 the New York State Health Commissioner announced that the landfill "constitutes a public nuisance and an extremely serious threat and danger to the health, safety and welfare of residents." Though the declaration made headlines around the world, the state was slow to take action. So was the federal government. Finally, in August 1980, when two EPA inspectors arrived to inspect the area, Gibbs and other residents held them "hostage" for several hours, demanding a commitment of action. Two days later President Jimmy Carter arrived and declared Love Canal a national disaster area. He agreed to federal purchase of all homes in the contaminated area and relocation of residents to safer neighborhoods.

Lois Gibbs could have moved to a new house in another Buffalo suburb and returned to the life she had left. Instead she moved to Vienna, Virginia, and started the Citizens' Clearing House for Hazardous Wastes (CCHW). By 1994 CCHW was a national organization providing management services, counsel, and financial support to thousands of ad hoc environmental organizations around the country.

From an environmental standpoint, Love Canal was not an isolated event. By 1978 the public had already witnessed dozens of environmental and public health disasters: a fire on the surface of the Cuyahoga River in Ohio, an enormous oil spill in Santa Barbara, the Kepone-poisoning of the wells of Hopewell, Virginia, the inadvertent mixture of a cancer-causing fire retardant with cattle feed in Michigan, the 17,000 containers of hazardous chemicals found in the "valley of the drums" near Louisville, the release of a dioxin cloud over Seveso, Italy, and a massive cluster of birth defects among infants in a Woburn, Massachusetts, neighborhood.

Alarming images of ruptured 55-gallon drums, flaming rivers, industrial ghost towns, deformed children, and "glow boys" in moonsuits diving into damaged nuclear containment vessels to receive a year's allowable radiation in a few seconds energized the American environmental imagination. The public began to understand that unregulated economic activity was poisoning them and that direct citizen action was essential. Lois Gibbs was a natural, almost inevitable phenomenon.

Then in March of 1979, as if to seal her fate, the Unit 2 reactor at the Three-Mile Island nuclear power plant suffered a loss-of-coolant failure. Two days later Pennsylvania Governor Richard Thornburgh issued an evacuation advisory for all pregnant women and preschool children living within a five-mile radius of the plant—about half the population residing in a fifteen-mile radius. The timing could not have been more propitious for the feisty barefoot epidemiologist and freelance environmental advocate: Gibbs was in Washington lobbying for the evacuation of women and children from her own contaminated community.

During the preceding 10 years there had been isolated protests against environmental polluters, and polls taken toward the end of the decade indicated that millions of Americans sensed they were at risk from chemical and nuclear exposures.[2] But Love Canal, an all-American, white, middle-class suburb—with its poisoned children and irate mothers—struck a chord in the American imagination that sparked a public demand for enforcement of the Resource Conservation and Recovery Act (RCRA) and other toxic-control legislation. Without social protests to draw national media attention, there had been no real incentive for EPA regulators to protect working-class neighborhoods. Lois Gibbs changed all that. A dozen years later sociologist Andrew Szasz called Love Canal "the grain of sand around which the perceptual pearl would begin to form."[3]

The pearl, of course, was a populist mass movement against toxic wastes. The American environmental movement would never be the same. Environmental populism had arrived, and the entire movement would be less polite for a while. A new class of activist—the angry mother—had been created. She was angry not only at the polluters in her community but, as often as not, at the mainstream environmental operatives in Washington who, by intervening with a compromise or negotiating over breakfast with some corporate executive, undermined efforts to remove a proven health threat from her community.

The environmental movement faced, simultaneously, danger and opportunity. The danger was that radical populism would fragment the movement into thousands of underfunded neighborhood organizations fighting isolated pitched battles against the petrochemical industry and its waste-management subsidiaries. The opportunity was the possibility of converting a stolid middle-class movement of liberal reformers into a diverse, militant, well-organized, and democratic political force to fight for environmental health at every level of society.

Love Canal remained on the front pages for almost two years. Toxic waste became a household nightmare and Lois Gibbs a national heroine. Many other women would emulate her. Sue Greer, Hazel Johnson, Cathy Hinds, Marie Sosa, Patsy Oliver, Kaye Kiker, Penny Newman, and Guadalupe Nuño transformed themselves from apolitical, diffident housewives who had never spoken in public to women who, faced with threats to their children and communities, found the courage to step forward and confidently lead the emerging environmental ad hocracy.

Prior to her involvement in the movement Sue Greer had never been more than three hours away from her home in Wheeler, Indiana. She now heads People Against Hazardous Landfill Sites (PAHLS), a powerful statewide coalition. Kaye Kiker's first response on hearing that Chemical Waste Management, Inc. (a subsidiary of WMX) was dumping hazardous waste into a nearby lot owned by Governor George Wallace's son was to pray. Today Kiker runs Alabamians for a Clean Environment, a nationally recognized model for statewide anti-toxic organizing. Cathy Hinds of Gray, Maine, lost most of the feeling in her arms and legs and then her newborn son before she became an activist. She is now director of the Military Toxics Campaign, a nationwide project to force the Department of Defense—the largest single source of toxic pollution in the country—to clean up after itself.

"The women in the environmental justice movement have made it a movement known for its irreverence for the powerful, its willingness to take strong positions of principle, its unending persistence, and its impatience with token solutions." says Penny Newman, former school teacher and founder of Concerned Citizens in Action. The group formed in 1979 to push for cleanup of the Stringfellow Acid Company pits in Glen

Avon, California. "These women have refused to play by the old rules and have been the ones to insist that enough is enough. It has been women, for so long shut out of decisions affecting themselves and their children, who have stepped forward to demand a say."[4]

American grassroots environmentalism now contains thousands of regional and local associations that sprang up in the wake of Love Canal to confront environmental insults—a toxic waste dump in rural Arkansas, an ocean incinerator off the coast of Texas, a pesticide sprayer in California's Central Valley, a refinery in New Jersey, or a railroad carrying hot nuclear waste through the small towns of Wyoming. When their efforts began to succeed, a public relations executive from the incinerator industry dubbed the ad hoc environmental organizations NIMBYs (Not-In-My-Back-Yard).

The idea was to discredit their motives and suggest that fighting to protect one's health, family, and neighborhood was some kind of moral defect or, worse, a social disease. H. Lanier Hickman, Jr, executive vice president of Governmental Refuse Collection and Disposal Association, a major promoter of waste incineration, called the NIMBY syndrome "a public health problem of the first order. It is a recurring mental illness that continues to infect the public." Hickman's solution was "a campaign to wipe out this disease." Others saw in NIMBYism the threat of anarchy. "More than a century ago de Tocqueville warned us of may be [sic] too much democracy in America," warned Calvin Brunner, a consultant to the waste industry. "Because everyone felt equal to everyone else, he projected that this would eventually lead to anarchy. . . . Is it possible that the NIMBYists will play a large part in proving de Tocqueville right in his assertions about democracy being [an] untenable form of government?"[5]

The NIMBY nickname stuck but did nothing to inhibit the movement. In fact, many grassroots activists now proudly refer to their organizations as NIMBYs.

By 1983 federal Superfund legislation was proving incapable of cleaning up the 30,000 abandoned hazardous-waste sites scattered around the United States, and Ronald Reagan had effectively eviscerated the EPA's enforcement capabilities. As far as toxics were concerned the mainstream movement seemed paralyzed. In response, John O'Connor, a young activist from Boston, founded the National Campaign Against Toxic Hazards, later renamed the National Toxics Campaign Fund (NTCF).

Like CCHW, NTCF became a national service organization for thousands of grassroots anti-toxic NIMBYs. NTCF pioneered a process called "good neighbor negotiations," in which anti-toxic groups and community officials assist local industry to implement pollution-prevention measures. O'Connor also established a Citizens' Laboratory to test air, water, and soil samples for the presence of toxins. The laboratory charges local communities approximately one-tenth the going rate at commercial labs. Like CCHW, the collective force of the NTCF coalition became far greater than the sum of its parts; although an integral part of the grassroots movement, it became a powerful national organization. In 1993 NTCF was disbanded because of internal disagreements. By then, it had greatly enhanced communications among thousands of NIMBYs and brought a much-needed scientific dimension to anti-toxics populism.[6]

Support, advice, and factual testimony have also been provided to ad hoc neighborhood organizations by the Clean Water Action Project and the National Coalition Against Misuse of Pesticides. Temporary alliances have been formed between local committees and public interest research groups (PIRGs), and, occasionally, Greenpeace (the largest of all membership organizations). Although regional chapters of the Sierra Club and field offices of NRDC have occasionally joined grassroots efforts, the local organizations mostly operate independently of the nationals, some of whose officials deride them as "radicals" or "Hammas" (a reference to militant Palestinians).

Grassroots organizations use many of the same tactics as mainstream groups—principally lobbying and negotiation—but they come to the table with an indignation that only a victim can display. Their anger and desperation are real and justifiable. "Would you let me shoot into a crowd of 100,000 people and kill one of them?" asks Lois Gibbs. "No? Well, how come Dow Chemical can do it to us?" "It's not illegal to build an incinerator and it's not illegal to poison people," Gibbs shouts to another assemblage. "Poor people know that they need to organize and fight to win."[7] That's not the kind of rhetoric one would hear from an official of the Sierra Club or the National Audubon Society.

It is impossible to quantify the scope of the anti-toxics movement. It is simply too vast and disparate. One indicator of its activity is Lois Gibbs' annual count of the local organizations CCHW has assisted with anti-toxic campaigns. In 1984 it served 600 community groups; by 1988 the number approached 5,000 and will probably pass 10,000 by Earth Day 1995.[8] At its peak, NTCF served over 7,000 grassroots organizations providing sound scientific assays of air and ground water for thousands of community activists seeking scientifically reliable evidence of pollution. CCHW has sent organizers from Virginia to over 200 cities. Although regional offices have recently been closed down for budgetary reasons, Gibbs and her staff remain in constant telephone contact with grassroots anti-toxic organizations in almost every state in the union.[9] In conjunction with the National Toxics Campaign Fund, the Remote Access Chemical Hazards Electronic Library (RACHEL, as in Carson), and other computer networks, CCHW altered the face of American environmentalism.

Today, grassroots anti-toxic environmentalism is a far more serious threat to polluting industries than the mainstream environmental movement. Not only do local activists network, share tactics, and successfully block many dumpsites and industrial developments, they also stubbornly refuse to surrender or compromise. They simply cannot afford to. Their activities and success are gradually changing the acronym NIMBY (Not-In-My-Backyard) to NIABY—Not In *Anybody's* Backyard. (The name of CCHW's monthly newsletter is *Everyone's Backyard.*) The grassroots are sending waste-producing industries a message that their production budgets are going to increase steeply in the next few years.

For business leaders accustomed to the polite and accommodating style of mainstream environmental leaders the NIMBYs have created a completely new problem. It can be argued (though it is difficult to prove) that grassroots environmental resistance during the past five to ten years has stopped more direct pollution than all the nationals' litigation combined. What is beyond dispute is that anti-toxics activism has been pivotal in turning industry toward genuine waste reduction. The end result of plugging the toilet, Gibbs and the anti-toxic movement hope, will be to make hazardous-waste management so expensive that industries will devise ways to reduce the waste stream of their manufacturing processes. Those who do not, will not, or cannot, Gibbs says, must be forced to manage their own hazardous wastes rather than shipping them elsewhere for burial or incineration.

The environmental ad hocracy has been so effective that the problem of siting facilities has assumed major proportions for the waste-management industry. Neutralizing the grassroots movement has therefore become a priority of the petrochemical and synthetics establishments. "They are the most radicalized groups I've seen since Vietnam," says William Ruckelshaus, CEO of Browning-Ferris, a waste-management firm based in Houston. "They've been empowered by their own demands. They can block things. That's a negative power. But it's real power. Right or wrong you can't bull your way through that kind of opposition."[10]

Despite concerted public relations campaigns to portray Gibbs and her colleagues as "hysterical," "irresponsible," "fear-driven," "self-centered," and "ignorant dupes," industry has convinced no one but themselves.[11] "What they throw at me is that I am a single-issue person," says Gibbs. "Yeah, I am a single-issue person. I look at the issue of people being poisoned and it makes me mad, and I wonder why it doesn't make everybody mad. It's a moral issue and that is why we won't go away. Our aim is to change the discussion within the boardrooms of major corporations," she adds. "That's where we will win ultimately, not in government agencies or Congress. Our strategy is basically like plugging up the toilet—by stopping them from opening up new landfills, incinerators, deep-well injection systems and hazardous waste sites." She predicts that "in the boardrooms at some point there is going to be this discussion: 'Hey, ten years ago, our disposal costs were X and now they are multiplying and so is our liability and so are our public relations problems.' That's when real change will come. All they understand is profit and loss. When the cost is high enough corporations will decide to recycle wastes, reclaim materials, substitute non-toxics in their products and eventually change their processes of production."[12] In a strange way, Gibbs and her cohorts, who are so critical of third-wave environmentalism and market-based incentives, are using direct action to force a massive market incentive on American industry.

Moreover, it's working. The populist anti-toxics movement has expanded exponentially and successfully chased waste handlers from county to county and state to state. "As a result there has not been a single new hazardous waste site opened anywhere in the country in the last 10 years," asserts Gibbs. "Without passing any new laws or regulations, without getting into the debate, we have stopped the expansion of hazardous waste sites in this country."[13] In the spring of 1990 William Ruckelshaus announced that Browning-Ferris was getting out of the hazardous-waste business altogether. It was, he said, simply losing money. Numerous government and business studies of the subject acknowledge, as a 1990 EPA report accedes, that "difficulties in siting new facilities owing to intense public opposition have begun to create capacity shortages." The Office of Technology Assessment agrees (1991) that "environmental organizations, public interest groups, and grassroots organizations have made waste reduction a priority." It seems that the toilet is indeed being plugged!

As CCHW and other networking organizations have grown, so have the targets of anti-toxic actions—from dump sites to incinerator sites, deep-well injections, landfills, pesticides, radioactive hazards, military nerve gas, acid rain, Agent Orange, electronic pollution, microwave radiation, and, eventually, to global threats such as ozone depletion and global warming. Gradually, as toxics were found to contribute to habitat destruction, species extinction, and loss of wetlands, the anti-toxics agenda entered the terrain of the conservationists and the potential for a broad environmental coalition became real.[14] Grassroots leaders, aware that toxic waste has become a major export of the industrialized countries, have also begun networking with environmental nongovernmental organizations (NGOs) in the third world. They hope one day to extend their acronym to NOPE (Not On Planet Earth).[15]

At the heart of the grassroots movement—and what to this day distinguishes it from the mainstream movement—is its strong belief in the right of citizens to participate in environmental decision making. Community right-to-know laws, citizen-enforcement provisions in federal and state legislation, and local input in waste-cleanup methodology and siting decisions are all central to grassroots demands. Thus a major challenge for the emerging ad hocracy has been combating the antidemocratic processes that all too often have been supported by mainstream organizations purporting to serve the public.

The essence of the grassroots complaints is that the nationals have become arrogant, elitist, insensitive to local efforts, obsessed with access, and prone to value wildlife above human life. At best, they see the nationals as service organizations able to provide occasional legal or moral support to grassroots issues or projects; at worst, they are obstructionists. When mainstreams get involved in local campaigns, decisions are often made in Washington or at state capitals, where deals may be struck in private without local consultation and with little concern for local consequences.

Here is how Kate Crockett, a grassroots activist with the Southeast Alaska Conservation Council (SEACC) in Juneau describes an all-too-common scenario:

> An environmental problem arises locally. People work to address the problem, develop a solution—frequently legislative—and gather support. As soon as the bill reaches D.C.—or sometimes the state capital—a national group comes in and with lobbying power and money takes over the issue and cuts a deal to pass the bill. They get the credit in the media *and* with the foundations. The grassroots lose in two ways. They don't get credit, and they are not at the table. . . . The reality is that there is always something to deal away in order to pass a bill. The grassroots are capable of determining what to trade, but when the deal is cut without them they see what they have lost and they don't own the decision.[16]

A real-life example cited by Crockett and others occurred in 1986, when several national organizations in Washington compromised with the chemical industry on an amendment granting federal pre-emption of state pesticide laws. When grassroots activists complained, mainstream leaders shrugged them off, saying it was the best they could do and better than nothing. Lois Gibbs exploded. The amendment would have completely defused all the influence that members of her coalition had painstakingly developed at state and local levels. Jay Feldman, director of the National Coalition Against Misuse of Pesticides (NCAMP), concurred. "It was a dramatic step backwards. It cut off the power [that grassroots activists] were developing."[17] In this instance, NCAMP and other grassroots groups were able to block the bill, but not before Gibbs flew several local organizers into Washington, not only to lobby Congress but also to reason with mainstream environmental leaders. The nationals eventually got the message and now oppose state preemption.

Grassroots organizations and the nationals remain at odds, however, over what Gibbs and others perceive as the nationals' excessive accommodations with industry in Clean Air Act negotiations and over recent attempts to weaken the Delaney amendment. The amendment prohibits the presence of all carcinogens, as residues or as additives, in food products. Mainstream organizations are increasingly willing to allow trace levels.

It can get worse, as it did in North Carolina in 1987 when officials from the state Department of Agriculture, advocates for the agro-biotech industry, and a representative of the Environmental Defense Fund met to draft the North Carolina Genetically Engineered Organisms Act (and accompanying regulations). The law, a bill which took about a year to develop, explicitly precludes any meaningful public participation in the regulation of biotechnology by enjoining local municipalities from blocking the release of genetically engineered organisms in their counties. The EDF representative in North Carolina, Steve Levitas, approved this precedent-setting "exclusion principle." EDF national leaders, while acknowledging that they have compromised the principle of public participation, argue that the best interests of the environment are served by the bill. As grassroots environmentalists read it, however, EDF aided and abetted the enemy and guaranteed the unchallenged commercial introduction of genetically engineered organisms.

Even worse, according to grassroots activists, is the fact that a national environmental organization had traded away community rights for a few dubious short-term gains. A consistent standard of public-interest advocacy says that lawyers should never trade away the rights of people affected by a statute or policy. The civil rights movement certainly never traded rights. They won them. There is no reason why the environmental movement should be trading basic human rights for access or other expediencies, particularly when such a dangerous precedent is established. Grassroots leaders would prefer that mainstream organizations become part of an alliance that refused under any circumstances to compromise with polluters. And when the environmental movement is approached by industry to "negotiate," local representatives ask only that they be invited to sit at the table.

In 1992 EDF again undermined the efforts of grassroots enviros, this time over the issue of ocean dumping off the coast of New Jersey. Clean Ocean Action (COA), a bicoastal coalition of 170 organizations, had for several years been fighting the practice of dumping sewage sludge, acidic waste, creosote ash, municipal garbage, and dredged mud at eight ocean sites along the northeast coast. By confronting the issue, dump site by dump site, with a creative mixture of litigation, political pressure, and direct action, COA had by 1992 closed down seven of eight existing off-shore dumps in the New York/New Jersey region. The only one left was the mud dump six miles outside the New York harbor. "The mud dump is the big one," according to COA Director Cindy Zipf. It receives all the dredging from rivers and channels in around the ports of New York and New Jersey. "The sediment on the bottom of those ports contains a legacy of 200 years of industrial waste," explains Zipf. "The Passaic River bottom, for example, contains the highest concentration of dioxins of any body of water in the country. But dioxin is only one compound. There are 51 other registered pollutants released directly into the water column at the mud dump. Fishermen are very upset about that."[18]

In 1992 COA joined United Fishermen, the American Littoral Society, the Confederation of Charter Boat Captains, and others in filing suit in Federal District Court against EPA Director Carol Browner, the Army Corps of Engineers, assorted dredging companies, and others responsible for the mud dump. "When we filed our complaint, 17 lawyers turned up to represent their side," said Zipf. "We had two. One pro bono." COA had invited Jim Tripp of the New York office of EDF to join them, but Tripp declined. "He said they didn't think it was a good case," recalls Zipf.

To everyone's surprise Judge Dickinson Debovoise called the government attorneys to the bench and excoriated them and the behavior of their clients. "As a citizen I am outraged," he said. "The plaintiffs know the law better than you do." Debovoise ordered the parties to mediate and assigned a mediator. Again EDF was invited to assist in the case, and again it declined. Mediation began a few weeks later, and by all indications things were leaning in favor of COA and its co-plaintiffs. "We were winning," says Zipf.

Then on March 31, 1994, without warning, EDF's Jim Tripp sent an eight-page memorandum to all parties involved in the mediation. The memo proposed a resolution favoring the position taken by the government. Because ports require periodic dredging, it argued, some place had to be found for the mud while a preferable method of disposal was developed. The mud dump should not, therefore, be closed until a suitable, cost-effective technology for disposal was found. This was basically the defendants' position. The plaintiffs' position, which until then was prevailing, insisted that no alternative technologies would be explored or developed until an early date for closure of the dump was set. Tripp's memo set the suit on a completely new course. "That memo completely undercut our position," says Zipf. "When I read it I felt like I had been kicked in the gut." Eventually the plaintiffs persuaded Tripp that his position was too generous to government agencies and the dredgers. By then, however, the mediator and the judge had reconsidered their own analysis and moved much closer to the defendants' position.

The tendency of some national organizations to embrace issues as if they were their own and even take undeserved credit for grassroots triumphs further exacerbates grassroots enmity. Again, the most revealing example of this tendency involved the Environmental Defense Fund, which, in 1992, agreed to enter negotiations with McDonald's over the use of styrofoam carryout containers. The initiative was taken by McDonald's after a nationwide consumer campaign against McDonald's led by Penny Newman of Concerned Neighbors in Action (Los Angeles) and Lois Gibbs at the Citizens' Clearing House for Hazardous Wastes. The company planned to place small incinerators behind every outlet to burn the containers (so they wouldn't fill up landfills). The campaign made the clamshell-shaped containers a symbol of wasteful throwaway packaging in the fast-food industry as Newman and Gibbs persuaded children across the country to send their used clamshells to McDonald's headquarters. Every day for months on end, McDonald's corporate mailroom in Oak Brook, Illinois, was inundated with thousands of smelly packages postmarked from every state in the union.

In desperation McDonald's President Edward Rensi called EDF's executive director, Fred Krupp. At Krupp's suggestion, a joint task force was formed. Neither Gibbs nor Newman were invited to join. The task force studied McDonald's waste stream and reached an agreement whereby McDonald's gradually would switch over to coated paper containers. (They are only a slight ecological improvement over styrofoam. Some believe them to be worse, because the bleached paper contains little or no recycled content.) The agreement fell far short of the expectations of Gibbs, Newman, and other grassroots activists.

In *The Nation,* journalist Kirkpatrick Sale generously declared it "an ambivalent success." EDF and McDonald's each sent out press releases declaring victory and taking credit for environmental protection. "It proves that McDonald's recognizes that the future is green," said Krupp. For McDonald's general counsel Shelby Yarrow it was a sweet deal. He found Krupp and EDF eminently reasonable: "I knew they wouldn't try to talk us into doing something if it was economically unreasonable," Yarrow told the *Wall Street Journal* after the deal was closed. Yarrow and Rensi circumvented CCHW and went directly to EDF because they knew "Fred wouldn't come in here and ask us to do Wedgwood china."[19]

EDF not only accepted a surrender but took credit for the McDonald's decision in future promotions, interviews, and newsletters. It never mentioned the indispensable role played by Penny Newman, Lois Gibbs, and aggressive consumer activism.

The unfortunate long-term consequence of such capitulations is not only that they undermine the tireless work of grassroots activists, but also that they allow companies like McDonald's to look a lot greener than they are. The corporate exploitation of such so-called win/win compromises has been relentless. Companies compete through paid and free media to "outgreen" one another. That, of course, is life on the corporate food chain; it's predictable and understandable. But the environmentalists' complicity, and their own PR-driven tendency to turn compromise into false triumph, illuminates the desperation and impending moral crisis of the mainstream organizations.

An Enduring Legacy of Neglect

The environmental movement has not been practicing one of the laws of nature: strength in diversity.

John Cook, Environmental Careers Organization

Come Sunday morning there's going to be a new environmental movement.

Reverend Ben Chavis, President, NAACP, Saturday, October 24, 1991

Although he may not have thought of himself as one, Martin Luther King was something of an environmentalist. When he traveled to Memphis that fateful day in 1968, he did so to protest the working conditions of trash collectors, mostly black men whose complaints included exposure to hazardous wastes. The issues before him that day were not defined as environmental, nor was his primary concern ecological. But had King survived that trip and traveled to Warren County, North Carolina, in 1982 to protest the siting of a disposal plant for polychlorinated biphenyls (PCBs) in a largely black and very poor neighborhood, he would certainly have talked about "environmental racism" and "environmental justice."

During the early years of the movement, in an understandable attempt to build the broadest possible constituency, environmentalists often described the issue as one that affected everyone equally. We all live in the same biosphere, said the gospel, breathing the same thin layer of air, eating food grown in the same soil. Our water is drawn from the same aquifers, and acid rain falls on the estates of the rich as forcefully as on the ghettos of the poor. On closer examination, however, massive inequities in environmental degradations and injustice in the policies used to correct them became evident.

While created equal, all Americans were not, as things turned out, being poisoned equally. The 10,000 people living in Chicago's Altgeld Gardens discovered this fact all too painfully. Altgeld Gardens' night sky is lit by the vapor lamps of heavy industry within the three-mile radius of oil refineries, chemical plants, sewage treatment plants, steel mills, and smelters that ring the community. Scattered through that industrial "doughnut" are 50 landfills, half a dozen incinerators, and over 100 abandoned toxic dumps. Health studies of the community show that less than half the residents are in good health. One survey revealed that about half the pregnancies in the families studied ended in miscarriages, birth defects, or sickly infants. About 25 percent of the children suffer from pulmonary diseases. The most relevant statistic, however, is that almost all the citizens of Altgeld Gardens are African Americans.[20]

People of color throughout the United States had long suspected that industry targeted their neighborhoods for its dirtiest businesses. In 1982 it came time to prove their case. In North Carolina, the selection of Warren County for a massive toxic-waste dump led to public demonstrations that attracted national media and led to over 500 arrests. Among those arrested were the Reverend Benjamin Chavis, then executive director of the United Church of Christ's (UCC) Commission for Racial Justice;

Walter Fauntroy, a member of Congress from the District of Columbia, and other civil rights leaders from the south.

Harassing congressional representatives carries certain risks. Subsequently, at the request of Fauntroy, the U.S. General Accounting Office (GAO) conducted a study of recent hazardous-waste siting in the southeastern United States. The study paid particular attention to the racial and economic makeup of the communities where landfills, incinerators, and deep-well injections were being situated. The GAO found that African Americans comprised a majority of the population in three of the four areas studied. The population of Emelle, Alabama—site of the nation's largest hazardous-waste landfill (which receives toxics from over 45 states)—is 78.9 percent African American.[21]

The elders of the United Church of Christ wanted to know more. They initiated their own exhaustive study and in 1987 released a report, *Toxic Waste and Race: A National Report on the Racial and Socioeconomic Characteristics of Communities with Hazardous-Waste Sites*. The report substantiated the GAO's finding that communities of color throughout the country were disproportionately targeted for commercial hazardous-waste and uncontrolled toxic-waste dump sites. UCC researchers also found that race, not income, was the primary determinant in siting polluting facilities and that three out of the five largest commercial waste landfills (accounting for 40 percent of the nation's garbage) were located in black and Hispanic communities. Moreover, about 60 percent of all African and Hispanic Americans and approximately half of all Asian, Pacific Island, and Native Americans were living in communities with uncontrolled toxic-waste sites. For those who believed the findings were a statistical fluke, the UCC report concluded that "the possibility that these patterns resulted by chance is virtually impossible, strongly suggesting that some underlying factors, which are related to race, played a role in the location of commercial hazardous waste facilities."[22]

An "underlying factor" of a community's ability to resist becoming the site of hazardous disposal was uncovered in a 1984 study performed for the California Waste Management Board by Cerrell Associates, a consulting firm for government planners. *Political Difficulties Facing Waste-to-Energy Conversion Plant Sitings,* the "Cerrell Report" as it became known, profiles neighborhoods most likely to mount a successful resistance: "All socio-economic groupings tend to resent nearby siting of major facilities, but middle and upper socioeconomic strata possess better resources to effectuate their opposition. Middle and higher socioeconomic strata neighborhoods should not fall within the one mile and five mile radius of the proposed site."

The report, which became the handbook for site location in the toxic-waste industry, stops short of mentioning race. It does, however, list many characteristics of minorities in describing those "least likely to resist": they are disproportionately middle aged or older; live in communities of less than 10,000 population; have a high school education or less, low incomes, and blue-collar jobs; and are "not concerned with issues." Also said to be characteristic of the nonresistant group are people in "nature exploitative occupations—farming, ranching or mining"—those who are "not property owners or whose property is of modest value," and the "religious and politically conservative—people with a free market orientation, who don't lean toward a socialist-welfare state." Once a site fitting that general profile is located, the Cerrell Report advises, "decide, announce, defend."

In 1992 the *National Law Journal* in Washington published a comprehensive analysis of all environmental lawsuits settled during the previous seven years. The authors, Marianne Lavelle and Marcia Coyle, found that "penalties against pollution law violators in minority areas are lower than those imposed for violators in largely white areas." The average penalty imposed by courts for violations of the Resource Conservation and Recovery Act was $335,000 in white areas and $55,000 in minority areas. Moreover, under the Superfund cleanup program, it took 20 percent longer to place abandoned hazardous-waste sites in communities of color on the National Priority list than those in white neighborhoods. Such racial imbalances often occurred, the authors noted, "whether the community is wealthy or poor." The factor in such cases was clearly not poverty. It was race.[23]

Lavelle and Coyle also conducted an analysis of every residential toxic-waste site in the Superfund program. They found that the EPA took much longer to address hazards in minority communities than it did in white communities. Even when cleanup was ordered, the agency was likely to order a "containment" procedure in minority areas and permanent "treatment" in white areas.[24]

Studies of air-quality demographics have produced similar findings. While most Americans breathe polluted air, only 57 percent of whites reside in counties with federally substandard air quality, while 65 percent of blacks and 80 percent of Hispanics live in counties with similar or worse conditions. In addition, Lavelle and Coyle's study of 352 Clean Air Act cases found that the populations benefiting from enforcement of the act were 78.7 percent white, 14.2 percent black, and 8.2 percent Hispanic.[25]

Such findings have turned thousands of people of color into overnight environmentalists. It was not that they hadn't known or sensed that they lived closer to toxics than white folks, but now they had scientific proof of how and why things came to be that way. Civil rights leaders and activists were particularly incensed by the UCC's findings and responded by either joining a grassroots group or forming one of a host of new local environmental organizations. In two years the number of grassroots anti-toxic organizations served by the Citizen's Clearinghouse on Hazardous Wastes increased from 2,200 to over 7,000; more than half of the new ones were founded and staffed by people of color. Gibbs' felicitous choice for the Clearinghouse's motto is "People united for environmental justice."

The mass media often describes environmental justice as "the new environmental movement," but it isn't really new, particularly to its victims. The fight against environmental racism has been in existence for as long as the mainstream environmental movement, operating parallel to it for 25 or 30 years. Until recently, of course, it didn't refer to itself as *environmental*—in fact, it some circles, it still doesn't without carefully adding the word *justice*. Furthermore, until even more recently, mainstream environmentalists overlooked environmental injustices, even when brought to their attention in myriad forms.

In a collection of essays published for Earth Day 1970, Adam Walinsky, a Democratic candidate for attorney general of New York, wrote: "The environmental movement is in serious danger of ignoring the most serious threats to the environment of the United States today. . . . The movement is well on its way toward passing by the two groups which, even in the narrowest terms, face the worst environmental problems in the nation." Walinsky was speaking of the poor and of "men and women who work in plants and factories." Middle-class environmentalists and reform politicians, he said, were unable to understand the environmental plight of the working man and woman. "We do not share his [*sic*] troubled neighborhood or his difficult and dangerous job."[26]

By the mid-1980s, however, there were signs that industries that had been reluctant to consider waste reduction were being driven to it. Among the reasons cited by the National Research Council, the Office of Technology Assessment, and the EPA were "higher costs of disposal," "public opposition to hazardous waste facilities siting," and "grassroots organizations that had made waste reduction a priority." The NIMBYs—black, white, and brown—were winning, although the statistics of environmental racism remained staggering. Studies completed since 1990 indicate that:

• Close to 50 percent of all African-American infants tested for lead contamination still had blood levels higher that the U.S. Center for Disease Control's standard.

• According to an EPA study released in the spring of 1992, three out of four toxic waste dumps not in compliance with federal regulations are located in black and Hispanic neighborhoods.

• Communities in which incinerators are sited have 89 percent more people of color than the national average.[27]

• More than 15 million African Americans live in communities with one or more uncontrolled toxic-waste sites. Approximately half of all Asian Americans, Pacific Islanders, and Native Americans also live in communities with such sites.

• More than 200 million tons of radioactive waste lie in tailings piles on Indian reservations. The rate for cancers affecting the sex organs among Navajo teenagers is 17 times the national rate.

• Every year 300,000 farm laborers (mostly Hispanic) suffer pesticide-related illnesses and disorders. (The EPA stopped collecting data on pesticide poisoning on orders from the Reagan White House in 1980. During the entire Reagan era not one, single Hispanic was involved in federal decisions related to farm worker protection.)

• Pollution-induced asthma among inner city blacks is many times higher than the average for whites. It kills five times as many blacks as whites. Black urban male children are almost three times more likely to die of asthma than their white counterparts.

- A survey of mainstream environmental organizations found that one-third of them had no people of color on their staff, and over one-fifth had none on their boards. [28]

In 1990 the last finding prompted several civil rights leaders to send a letter charging them with "racist hiring practices" to the National Wildlife Federation, Friends of the Earth, the Wilderness Society, the National Audubon Society, the Sierra Club, the Sierra Club Legal Defense Fund, the Natural Resources Defense Council, and the Environmental Defense Fund. The first letter came from the Gulf Coast Tenant Leadership Development Project in New Orleans, and the second from the Southwest Organizing Project in Albuquerque, New Mexico. Both are led and staffed primarily by people of color and in recent years have become actively involved in environmental (mostly anti-toxic) activity. "Racism and the whiteness of the environmental movement is our Achilles heel," read the Gulf Coast letter. "You must know as well as we do that white organizations isolated from Third World communities can never build a movement." It was a stinging indictment for organizations that considered themselves progressive and generally enlightened. But the facts were undeniable. Very few minority people held meaningful jobs in the nationals, and even fewer were in professional or management positions.

The letters also called for a more equitable distribution of resources, a challenge to mainstream environmentalists working in communities of color to provide more than moral support and free advice to grassroots organizations fighting big industry and big government. Similar requests came from third-world environmentalists, who called for recognition of their sovereignty and asked that international organizations work with rather than against the interests of indigenous people.

Reactions to the letters varied, but the most common response was "we tried, but there was no one out there." Minority enviros were baffled. "It is crystal clear that diversity in nature is superior to homogeneity," wrote Dorceta Taylor, "why isn't it obvious to people advocating this position that such diversity is also desirable in human communities?

Why isn't it understood that diversity is necessary for this movement to grow and progress?"[29]

Some mainstreamers were also stunned. "It was like someone took a two-by-four and hit us on the side of the head," lamented then Sierra Club Director Michael Fischer. "It was a wakeup call. They were saying 'you are getting lazy.' " Some leaders were responsive. Catherine Verhoff, director of operations at NRDC in New York, warned that "if environmental groups don't heed these people—and this is a real populist grassroots movement—the major national groups are in real danger of becoming obsolete . . . We can go with them, work with them and all become part of one big movement or we can be run over by them."[30]

At about the same time, the Network for Environmental and Economic Justice wrote to Greenpeace, the National Toxics Campaign, and the Citizens' Clearing House for Hazardous Wastes, expressing deep appreciation for their support of grassroots struggles in communities of color. The letter pointed out, however, that their organizations were still led and controlled by whites and were thus more likely to advocate *for* rather that *with* communities of color. Since then Greenpeace has made a concerted effort to hire people of color at all levels; it has also earned additional support from the environmental justice community for organizing an international campaign to protest the export of mercury-contaminated waste to South Africa.

The reaction of most mainstream, Green Group organizations to the letters was to mount a "diversity offensive" and circulate grant proposals to foundations to fund it. "It does not take an enormous commitment for environmental and conservation groups to make small steps for big bucks under the banner of diversity," complained Robert Bullard, a sociology professor at the University of California, Riverside. "The true test of commitment to diversity is the extent that these organizations use their own funds to diversify. Community groups are now forced into a state of economic warfare with the Group of 10." Bullard also predicted that mainstream groups would resort to "the WHOM (We Have One Minority) plan and that foundations would fail to use their leverage to insist upon real diversity. "Appointing people of color to boards and staffs is the easy part," asserted Bullard, giving them paid jobs and "assuring that their voices are heard is another matter."[31]

Race and environment have been concurrent political issues on the continent since well before the American Revolution. The earliest conflicts between private and public land were fought between European settlers and Native Americans, who had sharply divergent views about agriculture, hunting, and land stewardship. The elimination of the bison was intended both to clear land for domestic cattle and to "destroy the Indian's commissary." General George Crook gave his cavalry troops orders to shoot all buffalo on sight so as to starve out the natives.[32] It worked. Millions of Indians were forced by starvation to leave the Plains.

After being pushed onto barren reservations by governments sympathetic to ranchers, miners, and loggers, Native Americans thought they had seen all the callous discrimination and insensitivity they could imagine. They were shocked then when, two centuries later, white environmentalists took positions that jeopardized their survival. This occurred recently over grazing and timber rights on American reservations, most of which, by treaty, are autonomous regions. Mountain Utes were particularly incensed when the National Wildlife Federation sued the Bureau of Land Management to make it stop the leasing of five canyons on the Ute reservation in southern Utah. The tribe, which had been grazing about 200 head of cattle in these canyons for generations, intervened on behalf of the BLM, which lost the suit. The Utes lost another piece of their livelihood.

They were already at odds with environmentalists, who for years had opposed a small water diversion from the Animas and La Plata Rivers to bring irrigation and drinking water to parched towns on reservations in Colorado and New Mexico. The diversion was part of a much larger project conceived by the Bureau of Reclamation, which environmentalists had quite properly been fighting for years. The mainstream, however, for reasons of their own, refused to exclude the small diversion the Utes requested. The most recent lawsuit over this project, filed by Friends of the La Plata (represented by the Sierra Club Legal Defense Fund) was filed to protect the endangered Colorado squaw fish. "For 100 years we did not have running water on this reservation," lamented Judy Knight Frank, a Mountain Ute leader. "Where were the environmentalists then? They weren't hollering about the terrible conditions for our children, but now suddenly the squaw fish is so important—more important than the Indian people, apparently."[33]

Kettleman City

In the world of waste management, Kettleman City, California, closely fit the Cerrell Report profile of an ideal site for a hazardous-waste facility. The area is 95 percent Hispanic, extremely poor, rural, undereducated, and economically unstable. It is already home to the largest hazardous-waste landfill west of the Mississippi, which is operated by the world's largest chemical-waste handler Chemical Waste Management Corporation, a division of WMX Technologies. Surely the citizens wouldn't mind another trapping of thc garbage business, thought Chem Waste officials. But when in 1989 the company sought permission to place an incinerator in the community Kettleman City, population 1,200, surprised the company and defied the findings of the Cerrell Report.

The EPA had already approved, without a public hearing, the expansion of Chem Waste's hazardous-waste landfill. The site became a major *cause célèbre* for the Southwest Network for Environmental and Economic Justice. This coalition, formed in 1990, brought together more than 50 grassroots organizations in Texas, Oklahoma, New Mexico, Colorado, Arizona, Nevada, and California. When news of the approval was released the network confronted the EPA directly—in Washington and in the Region IX offices in San Francisco, which they occupied until officials agreed to sit down with them. The campaign—designed to pressure EPA into enforcing environmental regulations on an equal basis in communities of color—worked.

The EPA moved to conduct health studies and, for the first time in its 20-year history, opened its decision-making processes to democratic participation. The Kettleman precedent spread throughout the country. Top EPA officials began meeting with environmental justice leaders around the country, and the agency set about employing people of color in professional and official positions.[34]

While they were campaigning against the EPA decision, Kettleman City organizers Ausensio Avila, Esperanza Maya, and Mary Lou Mares approached several national environmental organizations for help. Only the Sierra Club responded with a $2,000 contribution to El Pueblo para Aire y Agua Limpio (People for Clean Air and Water), the ad hoc organization formed to fight Chem Waste and the EPA. The only active support came from Greenpeace and the Albuquerque-based Southwest

Network for Environmental and Economic Justice, both of which sent trained specialists to work with El Pueblo.[35]

When El Pueblo sued the King County Board of Supervisors they were represented, not by an environmental law firm, but by California Rural Legal Assistance (CRLA), a public-advocacy law firm set up many years before to represent Mexican-American farmworkers. As the county supervisors had already approved the incinerator, CRLA's only recourse was to sue them in Superior Court. CRLA cited the incinerator's impact on air pollution and complained that the Environmental Impact Report for the project was written in English in an area where when 40 percent of the residents read only Spanish. The court ruled for the plaintiff in 1992, and the Kettleman City incinerator was stopped.

At the annual Land-Air-Water-Law Conference held in Eugene, Oregon, a year later, Esperanza Maya repeatedly turned to a representative from the EPA with whom she was sharing the panel and asked him: "Where were you? Where were you when we were fighting for our lives in Kettleman City?" The EPA official did not answer, but Maya made her point by simply asking the question in public. It was the first time in the four-year struggle against the pollution of her community that she or any of her colleagues had come face to face with an EPA official.

When asked "where were *you* at Kettleman City?" (by this author), Vic Sher, executive director of the Sierra Club Legal Defense fund, cited the problem of "Triage—too much to be done that was of merit." A representative of NRDC who asked not to be quoted by name said "We simply didn't think we could win." At the time she questioned "whether the legal work was worth the trouble. We weren't likely to succeed, and we didn't want to lose. Besides, NRDC is not opposed to *all* hazardous waste incinerators." Acknowledging that El Pueblo had won its battle against the county and Chem Waste, she acceded that "we might look at Kettleman City in a different light today."

In a private interview Sharon Duggan, an environmental attorney in private practice, bemoaned the mainstream's compulsion to win. "Only litigating when you are confident of winning is a limited strategy." Duggan cited the civil rights precedent: "Thurgood Marshall filed many civil rights suits he knew he would lose, simply to force the issue before the Supreme Court. Even if the highest court voted against Marshall, they were compelled to hear his arguments and deal with the issue of civil rights. In the end, the strategy of losing paid off."

The Summit

Not one major environmental organization or conservation organization can boast of significant Black, Hispanic or Native American membership.
Peter A. A. Berle, President, National Audubon Society

If there are temporal turning points in the history of American environmentalism, the period between October 25 to October 27, 1991, is certainly one of them. During those days the First National People of Color Environmental Leadership Summit was held in Washington D.C. "The Leadership Summit is not an independent event," read the preface to its proceedings, "but a significant and pivotal step in the crucial process whereby people of color are organizing themselves and their communities for self-determination and self-empowerment around the central issue of environmental justice. It is living testimony that no longer shall we allow others to define our people's future. It is our intention to build an effective multi-racial environmental movement with the capacity to transform the political landscape of the nation around these issues. The very survival of all communities is at stake." The official goal of the summit was to "reshape and redefine the American environmental movement."

Three hundred delegates from 50 states, Puerto Rico, the Marshall Islands, Central America, and Canada met the first day to set an agenda. None were white. The second day they were joined by 250 representatives of civil rights, community development, population, church, health, and mainstream environmental organizations. Summit organizer Dana Alston told the newcomers, most of them white, that the gathering had not been convened to oppose the racial majority or the environmental movement, "but rather to affirm our traditional connections to and respect for the natural world, and to speak for ourselves on some of the critical issues of our time. . . . The issues of the environment do not stand alone by themselves," Alston continued. "They are not narrowly defined. For people of color, the environment is woven into an overall framework and understanding of social, racial, and economic justice. The definitions that emerge from the movement for environmental justice are deeply rooted in culture and spirituality and encompass all aspects of daily life—where we live, work and play."

Alston was followed by Pat Bryant, director of the Gulf Coast Tenants Association, which has organized residential communities along Louisiana's "Cancer Alley," a 28-mile stretch of the Mississippi between Baton Rouge and New Orleans where scores of oil and chemical companies have built refining and manufacturing plants. Bryant pointed out that most of the neighborhoods along Cancer Alley were occupied by the direct descendants of former slaves and stated that "environmental organizations had to know this, and that this was genocide. [But when we] reached out to national organizations, they couldn't understand us." Charon Asetoyer, a South Dakota Sioux, described her people's defeat of attempts to site a landfill and medical-waste incinerator on a reservation where unemployment exceeds 80 percent. "Before they came to us," she said, "fifty-three other tribes had been approached by the same people. We seem to be easy targets."

Mainstream environmental leaders were invited to the summit. A few came. Michael Fischer, then executive director of the Sierra Club, confessed that the club had been "conspicuously missing from the battles for environmental justice. We regret that fact sincerely. We are here to reach across the table and build the bridge of partnership with all of you. We are not the enemy." Fischer cautioned that "the divide-and-conquer approach is one that the Reagan and Bush administrations have used all too successfully all too long. We know that it is in our enlightened self-interest to be fully involved in seeking environmental justice *or we risk becoming irrelevant*" (emphasis added).

John Adams came from NRDC and announced that he and his organization had been challenging the "disproportionate impacts on communities of color" for some time. He promised to address the colorless hue of his own staff. "The NRDC has come to realize that the issues of racism, poverty, and environmental degradation are intertwined and inseparable," he said. "The environmental movement cannot solve one without solving or attempting to solve the others." Adams and Fischer both expressed hope that white and nonwhite environmentalists could avoid racially divisive politics that play into the disruptive efforts of the conservative political movement. "You can't win this battle alone," Adams warned the gathering, evidently still not comfortable enough with the situation to say "we."

Adams and other leaders have made it a priority to hire nonwhite staff and professionals and have even been accused of raiding talent from the grassroots movement. Although some organizations have successfully created staffs that racially reflect the general population, in all nationals top management, professionals, and department heads are still mostly white.

What angered the environmental justice movement far more than the slow rate of integration was the speed with which mainstream organizations wrote and circulated "environmental justice" grant proposals after the People of Color Summit. Dana Alston seemed to foresee what would happen. She predicted at the Summit that "If resources become available from foundations and donors, we have to understand that environmental organizations might go out in the name of working with us and raise a lot of money while our organizations continue to struggle over every penny and every dollar." Some delegates were particularly incensed when only days later the National Wildlife Federation drafted and circulated to large foundations a $1.4-million proposal to work on environmental justice issues. The proposal, they complained, was drafted without adequate minority input or consultation with grassroots leadership. "To watch an organization with Waste Management on their board raising money for environmental justice was very difficult for us," according to Alston. "That company dumps some of the most hazardous materials known to science on the communities of people of color." Lois Gibbs too was distressed: "We have built this wonderful movement, and it's almost like being victimized again."[36]

Before they adjourned, delegates to the summit drafted and ratified 17 "Principles of Environmental Justice" (see appendix for full text). As if to reassure traditional environmentalists who believe newcomers to the movement are obsessed with toxics and don't care enough for the preservation of wild nature, Principle 1 reads: "Environmental justice affirms the sacredness of Mother Earth, ecological unity and the interdependence of all species, and the right to be free from ecological destruction."

After the summit environmental justice activists held several regional meetings throughout the country. The largest was convened by the Southern Organizing Committee in December 1992 in New Orleans.

Over 2,000 people attended and developed a communications network for southern activists and organizations. The network subsequently joined a nationwide structure of organizations that collaborate and exchange information, strategy, and tactics. Although the question is still widely debated among activists, the grassroots have made a concerted effort to avoid forming a centralized process or organization. According to Alston, "There is a related antipathy for national organizations" bred by experience with the traditional environmental movement. "The networks now emerging are self-consciously trying to balance the natural tensions between national and local groups over setting agenda, sharing resources and spreading leadership opportunities. In essence, we are trying to create democratic decision-making structures that rely on strong, participatory and representative boards."[37]

Since the summit, environment justice leaders have collaborated with members of the Congressional Black Caucus in the reauthorization processes for the Resources Recovery Act (1992) and the Superfund Act (1993). In 1992, also immediately after the summit, then-Senator Al Gore introduced the Environmental Justice Act, which is considered "a good start" by most grassroots activists, although it "misses the mark" in protecting communities of color from hazardous-waste siting. During the transition from Bush to Clinton administrations, Dr. Robert Bullard, then a professor of sociology at the University of California, Riverside, was invited to serve on the Clinton transition team—the first time that the grassroots movement has been represented in the process.

Eighteen months after the summit and three years after the letters to the traditional environmental organizations, the Environmental Careers Organization (ECO) of Boston released the results of an employment survey of 61 traditional environmental organizations. *Beyond the Green: Redefining and Diversifying the Environmental Movement* reported that three-quarters of the groups surveyed still did not have ethnically diverse staffs. "The environmental movement has not been practicing one of the laws of nature: strength in diversity," read the first sentence of the report. Even though more than half the organizations claimed diversity as one of their highest priorities, and more than 90 percent had taken some action to achieve it, one-third still employed no people of color as staff, and more than 20 percent had none on their boards. "Human diversity in the environmental field is essential to creating a better environment," said the report. "However, the 'traditional' environmental movement has been exclusive. . . . The shortage of people of color in [traditional] environmental groups exists despite the growing body of knowledge concerning the disproportionate impact of environmental hazards in communities of color."

After the report's release ECO and several mainstream leaders sent a letter to 800 environmental groups across the country. It was signed by John Cook of ECO, John Adams of NRDC, Deeohn Ferris, director of the environmental project of the Lawyer's Committee for Civil Rights Under Law, and Charles Lee, director of research at the United Church of Christ Commission for Racial Justice. "We who advocate for the value of biological diversity have not equally embraced the power of human diversity," said the letter.

"Justice" or "Equity" at the EPA

From the early days of its existence, mainstream environmentalists have complained publicly about EPA intransigence. On countless occasions they resorted to lawsuits to force the agency to enforce the laws as its mandate required. Never, however, did they confront the agency for the overt racism that has been repeatedly documented by leaders of the movement for environmental justice. The EPA has not only neglected to pursue environmental justice for all, it has actively intervened against the interests and well-being of minority communities. Only a few examples are needed to make the point.

- In 1982, the EPA argued in court against African-American community activists in Warren County, North Carolina, who protested the dumping of PCB-contaminated soil in their community.
- It had long been known that black urban children had disproportionately high blood levels of lead. Not, however, until studies confirmed in 1984 that suburban children were also affected by lead poisoning did EPA move against oil refiners to lower the lead content of gasoline.
- The EPA has consistently maintained a staff of no more than six people to work on pesticide spraying, which is primarily a threat to farm workers. Over 50 staff people are assigned to radon contamination, which is primarily a threat to middle-class homeowners.
- When middle-class whites were threatened by the toxic contamination of Love Canal and Times Beach they were relocated and provided with

new homes. In 1990, when a cancer cluster turned up in a federally funded housing project built for farm workers in McFarland, California, no one was relocated. The project, it turned out, was built over a former pesticide dump.[38]

• In the course of regulating dioxin discharges from pulp mills in the Northwest in 1992 the EPA set levels of allowable discharge based on a human consumption level of 6.5 grams of fish a day, the national average. EPA officials *knew* that Native Americans, Asian Americans, and other low-income people in the area consumed up to 150 grams of fish per day—creating a cancer risk of 8,600 per million among people living below the poverty line. The risk allowable by statute is 1 per million.

Historically the agency has deliberately avoided issues of race and class in its deliberations and actions. In a 1971 hearing before the U.S. Civil Rights Commission, the EPA's first director, William Ruckelshaus, testified that his was a technical and scientific agency that was neither mandated nor equipped to deal with questions of disproportionate environmental impact. Over the years complaints of racism and injustice at EPA have mounted, many of them internally generated. When African-American academics at Howard University and the University of Michigan organized a conference on the matter in July 1990, Director William Reilly attended. Shaken by what he heard, Reilly assured the congressional Black Caucus that he intended to establish a work group of high EPA officials "to study risk and low income communities." He became an outspoken advocate of environmental justice. At a January 1991 tree planting in honor of Martin Luther King, Jr. he promised "the poor and disadvantaged the same protection from lead poisoning, hazardous waste and pesticide contamination as the country's most fortunate citizens. . . . I have a dream," he concluded, "that one day America will be for all people a land of beautiful, for spacious skies, from sea to shining sea."

Subsequently Reilly was invited to attend and address the People of Color Leadership Summit. When he failed to respond, he was disinvited. Later he was asked again but declined. Four months later, on February 24, 1992, EPA released the long-awaited report on environmental equity. The word *equity* was carefully chosen to avoid the hostile repercussions and reactions the agency feared would by evoked by *racism* and *injustice*.

An internal memorandum about the report expressed concern that environmental justice could become "one of the most politically explosive environmental issues yet to emerge." The clear intent of the "environmental equity communication plan" was to separate environmental from traditional civil rights activists. "Our goal is to make the agency's substantial investment in environmental equity and cultural diversity an unmistakable matter of record with mainstream groups before activists enlist them in a campaign that could add the agency as a potential target." The memo warned about the "long simmering resentment in the people of color and Native American communities about environmental fairness" and urged action before "the fairness issue [reaches] a flashpoint—that state in the emotionally charged public controversy when activist groups finally succeed in persuading the more influential mainstream groups (civil rights organizations, unions, churches) to take ill-advised actions."[39]

Minority environmentalists were naturally upset by the memorandum's obsession with political flashpoints and its complete disregard for real environmental threats faced by communities of color. They were also incensed by the use of *equity* and the avoidance of the word *justice* throughout the report. "When you use the word 'equity' it suggests that if we all share the problem it's OK," complained Charles Lee, director of the United Church of Christ's Special Project on Toxic Injustice.

Under Administrator Carol Browner the word *justice* is used with relative comfort at EPA, and programs have been initiated in response to an executive order to address environmental racism in agency affairs and decisions.

Labor and the Environment

The environment is too serious a business to be left to environmentalists.
J. William Futrell, President, the Environmental Law Institute

For many people, work is the most dangerous environment.
Charles Noble

America spends most of its waking time at work, yet it has only been recently that traditional environmentalists have come to regard the workplace as part of the natural environment. Clean air has meant smogless suburbs and a clear view across the Grand Canyon. The air breathed by workers in a Massachusetts factory was someone else's problem, not really the concern of environmentalists. Nor did they recognize that factories that pumped toxics into the air, soil, and water of their surrounding communities—definitely the concern of environmentalists—were also pumping toxins into the bodies of workers. At the same time as environmentalists were practicing "benign neglect" of workers, job blackmail and the threat of plant closure was discouraging the latter from seeing themselves as victims of environmental pollution.

Occupational diseases (not accidents) kill about 60,000 American workers, and workplace conditions cause an estimated 350,000 new cases of serious illness every year.[40] Nevertheless, occupational toxins have generally been a marginal issue in both labor and environmental circles. This situation is changing in part because substances that poison workers are leaking from plants into local neighborhoods. In addition, environmental justice groups are effectively exploiting the issue and challenging the "prove harm" philosophy of regulation. This assumption—which demands proof of harm before any regulation is implemented—holds all chemical compounds to be innocent until proven guilty and all ecosystems and species capable of assimilating toxins.

It's difficult to say exactly where and when the environmental/occupational health nexus occurred, if indeed there was a moment when low-income wage earners realized they were disproportionately affected by toxics. There were certainly well-publicized episodes that provided evidence—Bhopal was one; the day male workers in a small chemical factory in Lathrop, California, learned that the chemical they were manufacturing (DBCP) was rendering them sterile was another. Perhaps it was the day workers at the Hanford, Washington, nuclear facility realized that they had been "nuked" by their own employer.

At each of those moments, working people became reluctant environmentalists and began to recognize their workplaces as part of the environment. When they came to this realization, the struggle for occupational health and safety became an adjunct, if not a fully integrated partner, of the environmental movement. Mainstream environmental leaders were not wild about accepting the new environmentalists, who forced them to confront the corporate sector on a new battleground. Often the conflicts involved companies whose executives sat on the boards of national organizations.

The initial response of the corporate community to blue-collar enviros was to pit them against traditional environmentalists. "Too often the only time community based environmentalists meet workers is when we are protesting against corporate practices and the workers are bussed into public hearings to advance the company's agenda," laments Penny Newman, "so that the company can orchestrate a conflict between workers and the community."[41]

If there is ever a rapprochement between labor and environmentalists it is unlikely to be between unions and the mainstream environmental movement, and more likely to be between grassroots environmentalists and the rank and file—organized and unorganized. Many laborers and their families are drawn to grassroots anti-toxics activism whereas few are attracted to conservationism.

Relations between labor and the environmental movement have always been a bit tense, but not consistently so. Recent signs of rapprochement occur as workers in many industries begin to realize that the alleged tradeoff between environmentalism and economics is simplistic and essentially invalid. It is true that the coal miners union picketed Capitol Hill protesting the Clean Air Act in 1990. But the same year auto workers struck Chrysler over health and safety conditions in their plants. "Striking to Live" read one picket sign. "No more cancer" said another. Men and women suffering unusually high rates of cancer and respiratory and heart diseases from "fugitive emissions" of toxic fumes inside General Motors' Lordstown, Ohio, factories formed Workers Against Toxic Chemicals (WATCH). Even mine workers in some western states and in Appalachia, who are normally at odds with conservationists they perceive as threatening their jobs, have worked closely with environmentalists to pressure mining companies to restore land despoiled by strip mining.

In Louisiana, the Oil, Chemical, and Atomic Workers union allied with local environmentalists to mobilize protests over the environmental impact of a BASF plant. Workers and enviros stood persistently side by side on picket lines until BASF ended its five-year lockout of the plant. Union members then voted to hire and pay an environmental scientist out of their own dues to monitor the plant. When the scientist told them the plant was producing carbon tetrachloride, which most environmentalists agree should be banned, BASF workers began studying ways to shift manufacturing to other products.

Over and over corporate executives tell labor leaders that environmental regulations will force them to close plants and relocate production outside the country. "Industry begins the battle with a captive army of workers whose livelihoods are in some way dependent on the production of toxics and who are predisposed to believe company claims that environmentalists are well-to-do, anti-working class crybabies," according to Eric Mann, director of the Labor Community Strategy Center in Los Angeles. "Workers may argue in turn that if life is reduced to a battle between one self-interested force (environmentalists) attempting to take their jobs, versus another self-interested force (corporate management) attempting to 'save' their jobs, then they have no other self-interested option but to side with corporate power."[42]

Even though corporate power still prevails in most such situations, in recent years more and more workers, union and non-union, have begun to realize that in the long run opting for job security may not be the wisest or healthiest choice. Worker environmentalism is a new, potentially powerful force with which corporate America must contend.

Population and Immigration

Our goal in the United States should be achieving domestic population stabilization.

Carl Pope, Executive Director, Sierra Club

To want to limit immigration [in America] just because that is where most Sierra Club members live, I find extremely embarrassing.

Ruth Gravanis, Sierra Club Activist

First the numbers. Projections of future population vary, depending on the study, but these are conservative. The present world's population of 5.5 billion includes almost 2 billion malnourished, diseased people with very short life expectancies. It's hard to imagine that ratio improving much if population continues to grow exponentially. Barring catastrophes greater than any we have known so far, by the year 2025—only one generation hence—there will be 8.4 billion people on the planet; by 2050, 9.8 billion. These projections assume that current birthrates will continue to decline. If, perchance, birthrates stabilize at current rates, world population could reach 12 billion by 2050, doubling in less than 60 years. In either case, the ecological carrying capacity of the planet will be seriously challenged, as will the prophecies of Robert Malthus, Paul Ehrlich, and the Club of Rome.

Ehrlich's landmark book *The Population Bomb* (1968) placed the "population crisis" near the top of the environmental agenda. In 1972 the Club of Rome, using computer models developed at the Massachusetts Institute of Technology, predicted that at current rates of population growth, industrialization, resource depletion, and pollution the planet's limits of growth would be reached in less than a century. "Limits of growth" became a new mantra of American environmentalism, and the concept sparked a debate over sustainablility that still rages. Scores of solutions focused mostly on control of population. Only social democrats like Barry Commoner, visionaries like Donnella Meadows, and anarchists like Murray Bookchin proposed controlling industrialization, resource depletion, and pollution.

When apocalyptic predictions of food and resource scarcity failed to materialize, environmentalists moved population down their agenda. It was, as much as anything, an act of avoidance. The American environmental movement has always been deeply troubled and somewhat divided over the issue of human population, which ignites controversies and lays treacherous traps for liberal, humanitarian activists, environmentalist or not. Many simply choose to avoid the subject, but somehow they are forced to confront it. Few environmentalists deny that population growth is a problem, and their ideas about what to do about it vary widely. Is population control in the southern hemisphere as meaningful or consequential as population control in the North? Is population in and of itself as environmentally destructive as neo-Malthusians say it is? Or is mass technology and wasteful consumption really the problem, as liberals and radicals suggest?

One mainstream organization, the National Audubon Society, covered all bases in a 1993 statement: "Population growth cannot be isolated from factors such as poverty, lack of education and health care, unjust land tenure policies and overconsumption of natural resources by the United States and other industrial countries. All of these factors must be addressed in our efforts to foster environmentally sustainable development at home and abroad."

NRDC leans away from a strictly Malthusian position and lays stronger blame on the North: "Only a *fraction* of global environmental problems are directly linked to rapid population growth in less-developed countries," reads a 1993 position statement. "The industrialized nations (comprising only 22 percent of the world's population) must take the lion's share of the blame for the most serious pollution related problems."[43]

"An issue of at least equal importance to population," according to former Senator Gaylord Nelson, speaking for the Wilderness Society, "is the absence of a pervasive conservation ethic in our culture." Nelson's statement illustrates the fact that few, if any, of the mainstream organizations are comfortable discussing the role of social injustice, poverty, the status of women, or high infant mortality in population growth. Instead, they blame cultural or religious values and emphasize family planning and the export of birth control technologies as solutions. The impact of southern hemispheric population on the United States is of greater interest than its effect on the South itself. When debating Paul Ehrlich's $I = PAT$ formula (Impact = Population × Affluence × Technology), American environmentalists tend to believe that P is a much stronger determinant of environmental impact than A or T.[44]

Because the history of population control contains a legacy of racism and proposals by putative enviros such as Garrett Hardin that border on genocide, some mainstream environmental organizations have been understandably wary of taking stronger positions on the subject. Nevertheless, trustees, directors, and some active rank-and-file members of relatively conservative organizations like Sierra Club and the National Wildlife Federation, which tend to be somewhat nationalistic in their outlook, are beginning to push their organizations to take less ambivalent positions on population. Few, to date, have spoken out on the politically loaded matter of immigration, which has become an inescapable subtext of the population question.

Some mainstream environmentalists have recently expressed anti-immigrant opinions that are embarrassing to the leadership and deeply troubling to non-white staff and members. A few groups have even joined or formed alliances with immigration-reform organizations like the Federation for American Immigration Reform (FAIR), a national organization that blames many of America's social ills on immigrants. Supported partly by the Pioneer Fund, a foundation with historical roots to American neo-Nazism and eugenics, FAIR attempts to make anti-immigration sentiments palatable by forming coalitions with environmental organizations. National Audubon and EDF have both signed joint statements with FAIR, and prominent population activists in the Sierra Club are also active in FAIR (although it is uncertain which organization they joined first).

FAIR has some affiliations that are particularly disturbing to American minorities, especially Hispanics. Founded in 1979 by "English-only" booster John Tanton, a former member of the Sierra Club and president of Zero Population Group, FAIR emphasizes the "ecological" along with the "cultural" hazards of liberal immigration policies. Tanton warns of a "Latin onslaught" that will bring with it, he believes, such regrettable Latin American characteristics as Catholic hegemony, political corruption, and high birth rates. "Will the present majority peaceably hand over its political power to a group that is simply more fertile?" Tanton asks, cutting to the quick. "As whites see their power and control over their lives declining will they simply go quietly into the night, or will there be an explosion?"[45]

As a democratic organization with an array of chapters and special-interest committees, the Sierra Club has had to grapple with immigration policy in broad daylight. Unlike most environmental nationals, the club has a stated policy on immigration. It says in part that "all regions of the world must reach a balance between their populations and resources. Developing countries need to enlarge opportunities for their own residents, thus increasing well-being, eventually lessening population growth rates, and reducing the pressure to emigrate." Lest readers of their position fear that the club is laying blame or responsibility solely on the third world, the club's policy also states that "the Sierra Club urges Congress to conduct a thorough examination of U.S. immigration laws, policies and practices."

Immigration is a particularly difficult issue for the Sierra Club because of its past racial policy. In 1959 David Brower, representing the

San Francisco chapter, proposed a resolution to include "the four recognized colors" in the club's membership. The resolution failed. Again in 1973 the membership voted 3-to-1 against a proposal to involve the club in programs with racial minorities and the urban poor.[46] Today there is one African American on the board, a sprinkling of minorities on the staff, and a few members of color in the rank and file. Some of the latter are quite vocal in their insistence that the club involve itself in the emerging struggle for environmental justice. History, however, makes it difficult for the club's leaders to deal with population and immigration without drawing attention to its own past.

In 1992, in a possible attempt to redeem the club, at its 100th anniversary celebration in Harpers Ferry, Virginia, Executive Director Michael Fischer made an impassioned plea to turn the club over completely to minorities. (For full text of Fischer's address, see appendix.) There was polite applause, and members sang a verse or two of "John Brown's Body," but no motions were introduced. Fischer is no longer with the club. "The principal threat button that I pushed when I made that speech," he told me two years later, "was not cultural diversity, it was class diversity. National Sierra Club members just couldn't imagine themselves sitting down and talking to blue-collar or lower-income people."[47]

The Sierra Club has two committees that frequently clash over the issue of immigration. Frank Orem, an active member of FAIR, is chair of the club's national population committee. He supports a 1991 FAIR resolution, which he himself placed before his chapter's executive committee. It reads: "the U.S. should sustain replacement level fertility (2.1 children per family); the U.S. government should enact legislation establishing an all-inclusive legal immigration ceiling set at replacement level (i.e., immigration equals emigration); and the Sierra Club Bay Chapter should advocate legislation to implement such a policy." The resolution failed for lack of a second. But that hasn't stopped Orem. He has brought similar proposals before the national Population Committee and the full board, both of which are predominantly Anglo.

Countering Orem are representatives of the club's minuscule nonwhite membership and the Ethic and Cultural Diversity Task Force chaired by Chinese American Vivian Li. When the FAIR resolution was introduced neither Li nor her task force were invited to join the discourse. "I cannot begin to describe the anger and rage of many club leaders and members over the proposed policies, particularly the section on immigration," Li wrote to Orem. "Many members and leaders believe that the policy is ill-conceived, insensitive and racist, and will greatly damage the club's ability to become a more diverse and inclusive organization."

Another club member Norman La Force believes that American citizens need to alter their own lifestyles before blaming immigrants for environmental degradation. "We can turn people away at the borders," La Force says, "but unless we change our consumption patterns we will still use up the world's resources at a disastrous rate." Environmentalists whose approach is less nationalistic realize that environmentalism doesn't stop at national borders. They tend to be global in perspective and believe that anti-immigration, while doing nothing to solve global environmental problems, exacerbates racism.

In some form or other the issue comes before the Sierra Club almost every time they meet. Resolutions are introduced from all sides followed by acrimonious shouting matches that lead nowhere and alienate the very few racial minorities in the club. In May of 1994, in the midst of a heated debate a motion to table the motion until the next meeting was introduced. It passed.

Some foundations have complicated the issue further by offering large sums to consider the issue in worldwide perspective. In 1992 the Pew Charitable Trust created a program of Global Stewardship grants designed to "move the United States toward a position of leadership in . . . address[ing] problems associated with the worldwide interaction of population growth, wasteful and unsustainable consumption of resources, and deterioration of the natural environment." Nine environmental organizations applied for and received Pew grants for population programs.[48]

Alan Weeden, who serves on the National Audubon board and the national Population Committee of the Sierra Club, is also a trustee of the Frank Weeden Foundation, which funds primarily immigration and population programs (among them FAIR). In 1990 the Weeden Foundation granted $275,000 to the Sierra Club to set up a population program. Grants of $50,000 to $60,000 have been made to the project every year since. In fact, population advocacy is now the highest-funded program in the club's budget.

Weeden himself expresses a sincere intention "to treat the issue with sympathy, compassion and sensitivity to matters of race and color." That turns out to be an exceedingly difficult task for predominantly white clubs and organizations with enduring legacies of racism. For affluent white members with 2.1 children and investment portfolios containing stocks and bonds of companies profiting from resource-dependent, energy-intensive economies, population control remains the easiest fix. That doesn't, however, make immigration a safe subject; no matter what position one takes on either population or immigration the specter of racism is close at hand in a public discussion. Under these circumstances, the best, and most polite thing to do is to table all motions—a decision difficult to explain to foundations hoping for some action from their grants.

Communities of color are naturally sensitive to any discussion of population and immigration, particularly when the language of conversation is so often loaded with insect imagery ("swarms" and "hordes") and drowning metaphors (a "flood" of immigrants "pouring" over the border). Particularly disturbing to people of color are the observations and nostrums of some reputedly progressive enviros, notably Garrett Hardin and Earth First! co-founder David Foreman. Both men have introduced disturbing notions of social triage that to some observers border on genocide—if not active genocide, passive genocide. Foreman, at one point said that "if the human race should go extinct, I for one would not shed any tears" and suggested sending Mexicans home with rifles. He has since repudiated some of his less kindly views. Hardin has not, holding tight to his "Tragedy of the Commons" postulations that "distributional justice is a luxury that cannot be afforded by a country in which population overwhelms the resource base." Hardin asks, "How can we help a foreign country to escape overpopulation? Clearly the worst thing we can do is send food. . . . Atomic bombs would be kinder. For a few moments the misery would be acute but it would soon come to an end for most of the people, leaving a few survivors to suffer thereafter."[49]

Northern Hubris

That environmental justice has become an international issue was clearly evident at the 1992 United Conference on Economic Development in Rio di Janeiro. The United States delegation embarrassed itself and American environmentalists: first by helping eviscerate the global-warming treaty; then by declaring the third world's forests to be "sinks" for first world's carbon emissions; and finally by openly protecting multinational corporations from environmental blame. In addition, it totally ignored the issue of nuclear testing and blocked adequate funding for the few environmental remedies it did support. Yet the harshest label President Bush acquired from the mainstream American movement for his equivocal behavior in Rio was "ecowimp." "We're shocked, simply shocked" was the strongest response.

No one who attended the nongovernmental organizations' (NGO) preparatory meetings in New York would have been surprised by the behavior of American environmentalists in Rio. There and in Rio third-world delegates and observers found them imperious and insensitive. "We don't want to be lectured as to what we should do, unless it is done in a cooperative and democratic way," said Indian delegate Mani Shankar. "I am not about to go to my people and tell them they must face *more* deprivation because some lady in Maine is fretting over the cutting of a tree or because some chap in San Francisco wants to drive his Volvo in better conscience. We can sit down and talk when we realize that one job in Cincinnati is not one bit more important than one job in New Delhi."[50]

There were demonstrations every day in Rio, organized and conducted by northern hemisphere enviros. One *Journal do Brasil* headline was telling: "Americans and Europeans March in Defense of Animals, Forests, and Ecology." The subhead completed the story. "Brazilians protest hunger, poverty, and oppression."

Even so, Philippine environmentalist Maximo Kalaw believes that by the end of the conference American enviros at Rio began to understand the essential differences between northern and southern environmentalists. "There was a gravitational pull on all the northern groups," he says, "moving them toward real political concern for the needs of the poor. A year or two ago these same groups answered with a blank stare when you asked them their position on economic issues. They are now coming to the realization that their work is not just management of conservation projects but rather the facilitation of real political action."[51]

Martin Khor Kok Peng, director of the Third World Network and an NGO delegate from Malaysia, is not so sanguine. Khor believes that "anything positive that comes as a result of the Earth Summit will be nullified by the reality of the new world economic order," which he finds "a cause for deep depression; a return to an era of more direct colonialism, a negation of the principles of sovereignty over natural resources." Khor places responsibility for reversing this scenario on environmental organizations of the North, cautioning that to do so they must first deal with their dependence on "elites, who never give up anything voluntarily. Northern activists will have to bring pressure to bear on their own governments to start solving these problems where they originate."[52]

Mainstream Response

At some point . . . the movement got bogged down in bureaucracy, politics and the ordinary—probably in a well-intentioned effort to include the common man.

Patricia Poore, Editor, Garbage Magazine

Some traditional environmentalists, and evidently the editor of *Garbage* magazine, are still uncomfortable with efforts to involve "ordinary" outsiders in environmental protection. They find the participants "common," the agenda radical, and the vocabulary, particularly words like *empowerment,* alienating. One founder of a major environmental law firm, who spoke on background, postulated that opportunism might be at work in the push for environmental justice. "The civil rights movement was running out of steam," he said, "so it hitched itself to the environmental movement, to give itself a lift. Attempts were made long ago by environmental leaders to tell civil rights organizations that they should care about toxics. But we were ignored." Admitting to a degree of cynicism and requesting, again, to be kept off the record, the same person later admitted that the opposite could as easily be true—that the environmental movement was running out of steam so hitched itself to the civil rights agenda. Whatever happened, the environmental and civil rights movements need each other, and both will be the stronger for whatever coalitions they can build.

Anti-toxics litigators, for example, have already discovered a tremendous advantage working with the environmental justice movement. "We are on the verge of busting open toxics," Vic Sher, executive director and lead attorney at the Sierra Club Legal Defense Fund, told me. "For 20 years the environmental community has been boxed in by a set of legal restraints that focus on deference to agencies that make expert decisions. What will blow this apart is the civil rights argument which applies a completely different model of judicial behavior and a completely different set of questions." Fighting toxic pollution with civil rights laws, Sher believes, could win far greater protection than any of the laws written solely to protect the environment. "People have the right not to be poisoned. The environmental bar has never looked at the issue that way. The entire movement has to move in this direction if it to retain its relevance to larger society."

Less than a year after the People of Color Summit, the Sierra Club hired John McCown, a civil rights activist from Mississippi. McCown's mission was to establish links between club members and poor black communities in the south or, in his own words, "to challenge the sheltered lives of middle-class, educated European Americans [who] avoid African Americans like the plague."

After McCown led a tour of Sierra Club officials through a neighborhood bordering a Superfund site in rural Columbia, Mississippi, Robert Cox, now club president, was overwhelmed. It was the thirtieth anniversary of Medgar Evers' assassination. "As surely as Medgar Evers was shot," said Cox, "the people of this community are being poisoned." Not all his colleagues agreed. "A high fat diet can do it too," quipped club paralegal Michel Klaes. It was a beginning.

When the subject of real support for Columbia came up, club leaders said they could help local organizations apply for money from foundations or lobby Congress to relocate families but could do little in the way of financial assistance. In fact, several mainstream organizations, the Sierra Club among them, sought financial assistance from foundations to start their own environmental justice programs. These moves incurred considerable resentment in the People of Color Network, who saw white, middle-class fundraisers applying for the very funds they needed to organize and expand their movement.

John McCown, for one, is unperturbed by the criticism leveled at his employer. "I strongly believe we are being tested here," he says. "For so long we have allowed industry to divide us on the basis of race and class. We're going to have to come together as brothers and sisters—as human beings—to stop this problem of environmental injustice. As John Muir said, there's a connectedness between all things. If people in Columbia, Mississippi, are suffering, then something is ailing European-Americans from North Carolina to California as well." Robert Bullard agrees, but cautions that mainstream organizations must share resources with environmental justice groups. "For there to be real progress in improving relations there must be true partnerships between mainstream and grassroots environmentalist organizations," he said. "They must be equal."[53]

In the meantime, minority enviros remain cautious and independent. At several demonstrations of non-white environmentalists around the country, picket signs reading "We speak for ourselves" have been prominently displayed. The phrase, according to Richard Moore of the Southwest Organizing Committee, is "the watchword of the movement."

Reconciliation

The only thing worse than fighting with your allies is fighting without them.
Winston Churchill

It will not be easy to reconcile the enormous differences, so deeply rooted in class, culture, and history, between environmentalists whose primary interests are wilderness and wildlife, and blue-collar environmentalists struggling for personal survival. Not only are their goals different, at times, but their style, philosophies, and world views are so divergent, even conflicting, that they can barely communicate. As William Greider puts it, "An environmentalist who graduated from an Ivy League law school is more likely to believe in the gradual perfectibility of the legal system, the need to legislate and litigate. However, if one lives on the 'wrong side of the tracks,' downwind from toxic industrial fumes, such activities look pointless and even threatening."[54] Two very different cultures have fallen together under the environmental umbrella, not by accident but by fate. The task of working together for a healthier community will be challenging for both. Although diplomatic attempts have been made to reconcile the cultural, political, and class differences between grassroots and mainstream environmentalists, many Green Group professionals still have difficulty believing that the air in a Los Angeles furniture factory is as environmentally significant as the air in the Grand Canyon.

Unfortunately, there are no "watering holes" in the environmental movement, laments Conservation Fund President Patrick Noonan. "There are no forums where volunteers and professionals can share ideas, where land-conservation specialists can meet solid-waste campaigners, where air quality can meet water quality. There is no place where politicians can find a constituency, no marketplace for funders, and, most importantly, no platform for a continuing dialogue between economic and environmental interests."[55]

As far back as 1979 an attempt was made to close the divide between the civil rights and environmental movements. It was part of Jimmy Carter's initiative to bring blacks and whites together over issues of common interest. The administration funded a meeting, which was co-hosted by the Sierra Club and the Urban League in Chicago. Accounts of the meeting vary, but no one calls it a success. The gathering came to an immediate impasse over nuclear power. The Urban League was for it. When environmentalists attempted to pass a resolution opposing nukes, they were stifled at the podium. Later enviros complained about the luncheon speaker, a public relations executive from Atlantic Richfield Corporation. When organizers of the conference refused to muzzle him, the enviros pulled the plug on the microphones. Some planners of the conference and most who attended still considered it an exciting first step in an essential dialogue. It would be years, however, before such a rapprochement was attempted again.

The second attempt to create a temporary watering hole for diverse enviros was made several years later by Lois Gibbs, who invited about 30 of her community activists to meet in Washington with G-10 lawyers and lobbyists. "It was hilarious," Gibbs remembers. "People from the grassroots were at one end of the room drinking Budweiser and smoking, while the environmentalists were at the other eating yogurt. We wanted to talk about victim compensation, they wanted to talk about 10 parts per billion and scientific uncertainty. A couple of times it was almost war. . . . We were hoping that by seeing these local folks, the people from the Big Ten would be more apt to support the grassroots position, but it didn't work that way. They went right on with the status quo position. The Big Ten approach is to ask: What can we support to achieve a legislative victory? Our approach is to ask: What is morally correct? We can't support something in order to win if we think it is morally wrong."[56]

The first real step toward reconciliation, it seems, must be semantic. The definition of *environment* has to be expanded to include factory interiors and poor peoples' kitchens. *Environmentalism* should encompass the civil rights of children and the human rights of the Ojibway. Leroy Jackson, Lois Gibbs, and Cesar Chavez should be described whenever possible as *environmentalists*. Once those definitions are accepted and understood, the groundwork for moral, conceptual, and philosophical change will be complete.

Robert Bullard believes that for reconciliation to occur the environmental justice movement must move beyond toxics, lead in drinking water, and disproportionate impacts. "We have to keep saying 'It's bigger than that. It's about environment *and* economics. Its about peace and social justice and civil rights and human rights. They're all part of the environment." If, in the course of achieving environmental justice, the output of toxic pollution is lessened, the ultimate benefit is for all life, not just human life. This fact should please even enviros whose sentiments lean primarily toward wildlife, biodiversity, and species extinction.

One hopeful sign that reconciliation is actually taking place is the fact that the best environmental voting record in Washington is now held by the 40 members of congressional Black Caucus. This is all the more surprising when one considers that the standards for environmental rectitude are set by the League of Conservation Voters, a creation of the mainstream movement that reflects the national's environmental values and imperatives. The Black Caucus scores an average 76 on the league's scale, while the overall Democratic party scores 70, and the Republicans 24. Representative John Lewis (D-GA) offers an explanation: "It's because of the communities we represent. Many have been victimized by individuals and firms that have very little regard for the environment." That is not to suggest that black congressional interests are parochial.

The caucus has also voted strongly in favor of wilderness preservation, endangered species, marine mammal protection, and, taking a greater risk with party leadership, it voted with the Sierra Club in opposing the North American Free Trade Agreement.

Conclusion

American environmentalists have never used their numbers to full advantage. As William Greider says, "its goals are regularly stalled and subverted by the governing system, even when it wins." In recent years the situation has worsened as mainstream enviros have ignored the hazards of coercive harmony and fallen for the "win-win" rhetoric of alternative dispute resolution (ADR). This is not to suggest that mediation should never be attempted or that harmony is in itself a bad thing. But there are few issues or species left in the environment that can be compromised without grave risk. Win-win may mean less poison from a specific site, but it still means poison. And many poisons, as biologists have discovered, have a way of finding their way into and concentrating in human and animal tissue.

Greider also points out that "if any of the major environmental groups were to realign their own policies with those positive energies emanating from the grassroots, they would necessarily have to rethink their own policy priorities and methods and listen respectfully to what these people from the communities are trying to say. Inevitably this would put at risk the environmentalist's good standing as 'reasonable' participants in Washington politics. But they would also discover a source of new political strength—the power that comes from real people."[57]

In such a realignment of the movement the mainstream organizations will not and should not disappear. A federal strategy and the three Ls—legislation, litigation and lobbying—will always be vital aspects of environmentalism. But they are only aspects. In the end, the value of mainstream organizations will be measured by how effectively they work with grassroots environmentalists during the final years of the twentieth century. There is tremendous talent and passion in the mainstream environmental movement and a lot of potential leadership, but too much of it is being spent or misspent on bureaucracy and specialization. Those talents and passions belong on the battlefield, not in the office.

To remain relevant and effective, traditional environmentalists will need to expand their concept of environment beyond aesthetics and beyond wilderness to include "the place you live, the place you work, the place you play," as so often repeated by environmentalists of color. As Dana Alston reminds us, environmental justice must be "seen through an overall framework of social, racial and economic justice, and the environment itself as just one piece in a whole linkage."[58]

7

Faded Green

In the environmental movement, our defeats are always final, our victories always provisional. What you save today can still be destroyed tomorrow.

José Lutzenberger, Gaia Fundacao, Brazil

Power *never* yields without a struggle.

Frederick Douglass

As the second decade since Earth Day 1 drew to a close the luster of the green movement was fading. National environmental organizations in particular were suffering serious financial and membership losses. After a brief shot of enthusiasm, marked by a small rise in membership around Earth Day 2 (1990), the nationals hit the skids again. In most groups direct mail and canvassing shortfalls led to sudden budget deficits and painful layoffs. During the golden years none of the mainstream organizations had built endowments adequate to tide them over. The Sierra Club was in its worst financial shape in 20 years and was forced to pare its payroll by 10 percent. The club's membership dropped from 630,000 to 500,000, and it ran a cumulative operating loss of $7 million between 1990 and 1994—in spite of a sixfold increase in foundation funding during the period. The National Wildlife Federation laid off 100 people, and the membership of Greenpeace—by then the largest environmental organization in the world—plunged to 800,000 from a 1990 peak of 2.5 million. At the Wilderness Society too a 30 percent membership falloff forced the society to close offices and lay off staff. Even old line hook-and-bullet clubs like Ducks Unlimited were suffering.

Interpretations of the decline vary. Recession, certainly a factor, was a popular explanation among fundraisers unwilling to blame themselves, and among movement leaders reluctant to admit that their influence in Washington had waned. They did not pass a single piece of meaningful federal legislation after 1993. Yet the crisis of the greens has lasted too far into the recent economic recovery for recession to be the sole cause.

The Clinton-Gore administration, or, as some call it, "the Gore effect" is certainly a factor. Both Democratic candidates seemed so committed to the environment that grantmakers and donors alike perceived less of a need for their philanthropy. Another problem was undoubtedly "list fatigue," a direct marketing side effect that results from the overuse of mailing lists—the same old names get the same old stories; eventually it all stops working. Another reality is even harder to face: "The movement has lost support because it has failed to demonstrate convincingly that its goals are compatible with the economic needs of the country," explains Roger Craver, a direct mail consultant to several national organizations.[1]

Moreover the environment is competing for funds with other pressing human needs like AIDS, mass starvation overseas, and homelessness at home. When nature and the human domain are seen as separate entities, people feel they must choose between them. "At a time when human problems are mushrooming, people are forced to choose between helping nature and helping human beings," wrote Daniel Lazare. "Ninety-nine out of a hundred will chose human beings." Lazare's estimate is a bit extreme. In fact about 3 percent of American philanthropy still goes to environmental causes, but less of it goes to the nationals. The latter have had to compensate for declining membership revenues with foundation grants, music concerts, and gimmicks like the National Audubon Society's offer of free membership to anyone who buys a stuffed toy named Nature Bear.

Whatever the reasons, the mainstream movement sees itself at a crossroad. Over the past two decades its organizations invested heavily in federal strategy and created a large professional cadre of lawyers and lobbyists, who remain convinced that environmental salvation lies within the Beltway. They lament that their members no longer share their enthusiasm for Washington solutions and that they have been kept outside the gates of power for too long.

Crashing the Gate

What started out like a love affair turned out to be date rape.
Jay Dee Hair, National Wildlife Federation

The Clinton administration is between a rock and a sponge.
Tim Hermack, Native Forest Council

For most of December 1992 the American mainstream environmental movement was in a state of near ecstasy. At every opportunity, the president elect was publicly bear-hugging the most environmentally sensitive vice president in American history. It was true that Clinton had an atrocious environmental record as governor of Arkansas and very few enviros supported him during the primaries.[2] But Albert Gore was his redemption.

At the Environmental Ball, held in Washington the night before the presidential inaugural, enviros had tears in their eyes. Bruce Babbitt, president of the League of Conservation Voters, was headed for the Department of the Interior and Carol Browner, Gore's former Senate aid, was on her way to the Environmental Protection Agency.[3] Bill Clinton had promised to elevate the EPA head to full cabinet status. A rear admiral at the Department of Energy was to be replaced by a woman, and retired Senator Tim Wirth seemed certain to land a top environmental job in the new administration.

Babbitt was particularly exciting to a conservation movement that had been battling with the likes of Walter Hickel, Donald Hodel, James Watt, and Manuel Lujan. Here was a son of the frontier, a supporter of the Endangered Species Act who said publicly that the West was ruled by "the lords of yesterday" who believed that grazing, mining, and timber laws were stuck in the nineteenth century. The fact that he was a multimillionaire whose family had made a fortune in coal and other natural resources somehow didn't faze environmentalists.

At the transition team office, Gus Speth—a founder of NRDC on loan from World Resources Institute to head the Natural Resource Cluster—was considering the résumés of dozens of environmental professionals anxious to ascend to public service. After 12 years of being "barbarians at the gate," this was comforting news for many enviros. The word was that Speth had designed a revitalized Council on Environmental Quality (CEQ), which he himself would head. Details were vague, but the new CEQ, if Speth got his way, would be, for the first time since Nixon established it in 1970, powerful enough to actually enforce the National Environmental Policy Act.

Even Beltway environmentalists who didn't land jobs gained something they prized almost as much—access. Access to the president would be assured through Al Gore, who had heightened his popularity among greens by announcing that Rachel Carson would "sit in on all important decisions of this administration." And, through the friend they all called "Gus," they would have access to the four key environmental agencies—Interior, EPA, Energy, and Agriculture. For the first time ever there would be a true champion of the environment and an environmental czar in the White House.

By the following winter the thrill was gone. The EPA was still not in the cabinet, and its budget had been cut by $217 million; the Delaney clause was under official assault; and the concept of negligible risk was embedded in the philosophy of regulation by executive order. CEQ had been shut down and replaced with the White House Office on Environmental Policy, a comparatively puny adjunct of the vice president's office headed by 30-year old Katy McGinty, another former Gore aide. McGinty—competent, bright, tireless, and environmentally committed—found herself competing for the president's attention with national economic counselors Lloyd Bentsen, Robert Rubin, Ron Brown, Roger Altman, and Mike McClarty. Gus Speth, who would surely have fared better in their midst, was appointed administrator of the United Nations Development Program. It was a consequential assignment to be sure, but it gave him minimal access to Clinton or the National Economic Council.

About two dozen environmentalists were hired directly from national environmental organizations and salted strategically throughout the new administration. They were assigned not only to obvious domains like EPA, Agriculture, and Interior but also to newly created environmental offices at State, the Office of Management and Budget, and the National Security Council. Most—like Rafe Pomerance, who moved from the World Resources Institute to the State Department; Mary Nichols, formerly at the Natural Resources Defense Council (NRDC), who went to EPA, and Wilderness Society President George Frampton, who was hired by Babbitt to head the Fish and Wildlife and Park services at Interior—are in sub-assistant deputy positions with some advisory but limited policy powers. Initially, that mattered less to mainstream leaders than the very fact that they were there. "I can't tell you how wonderful it is," declared National Audubon lobbyist Brock Evans, "to walk down the hall in the White House or a government agency and be greeted by your first name." One of those who greeted him was his former colleague and National Audubon lobbyist Brookes Yaeger, who went to Interior. When Yaeger found himself side by side with Rafe Pomerance as official U.S. representatives to a conference in Geneva, they grinned at each other and Yaeger whispered, "Can you believe this?"[4]

It's difficult to say exactly what Clinton's strategy was in positioning enviros where he did—if there was a strategy at all. No one on or close to the transition team seems to remember hearing one. But hiring William Reilly to run EPA certainly helped President Bush keep the Washington environmentalists at bay for a while. Back then even a veteran like the NRDC's John Adams had been taken in by it enough to say, after a pre-inaugural breakfast, that he was willing to give Bush a chance "to turn out to be a 1989 version of Teddy Roosevelt."[5]

Whatever Clinton's motive, it has served the administration better than he has served the environmental movement or the environment. The first-name euphoria allowed both Clinton and Gore to fudge most of their greenest campaign promises for a full year with barely a whisper of protest from the Washington environmental community.[6] Grassroots environmentalists—from Liverpool, Ohio, site of a massive garbage incinerator, to guardians of the northwest's ancient forests—howled at the betrayals, but few in Washington were listening.

Environmental leaders in Washington were being particularly careful not to offend the administration because they perceived it, in some degree, to be their only hope for reasonable access to government agencies where they had been cold-shouldered for 12 years. Getting rid of Bush and Quayle *was* a positive step for the environmental lobby, but in the same election six very strong environmental representatives and senators either lost their seats or retired from office. Four of them were the congressmen who had managed the critical floor fight to prevent passage of the badly compromised Montana Wilderness Bill. Senator Tim Wirth (D-CO) opted not to run and now holds a vital environmental position in the State Department, providing another reason for the environmental lobby to shift its attention from Congress to the administration.

In return for access and some passionate environmental rhetoric—generally delivered in safe venues like Earth Day celebrations and White House breakfast meetings—President Clinton earned a one-year free ride. Studious politicians know from experience that access defuses hostility in social movements. Organization leaders who have access to power are more docile and less likely to allow direct action or strong criticism from the ranks. As John Adams, CEO of NRDC points out, "If Clinton hadn't made those appointments, we'd be his enemy."[7]

Despite a vigorous transition campaign mounted on his behalf, Adams (who says he knew nothing of the campaign) remained "the outsider I have always been." But how far outside is someone who has breakfast with the vice president and whose recent employees are ensconced in offices next door to cabinet secretaries—as former NRDC lawyer Dan Reicher is to Hazel O'Leary at the Department of Energy? Reicher can call Adams from that office and get more attention than he ever did when he worked for him.

Through subalterns appointed from the midranks of NRDC, National Audubon, the Wilderness Society, and the Sierra Club, Clinton gained a sort of reverse access to the environmental movement. This not only allows him to "lobby" a movement of about 10 million voters but leaves the old guard safely at the top of the movement. And, by not elevating the testier young enviros to high-profile positions, Clinton shields himself from attack after making environmentally harmful policy decisions—of which there have been many.

Measures of Betrayal

Almost immediately upon taking office, EPA Chief Carol Browner announced she intended to re-evaluate the Delaney amendment to the Food and Drug Act. Delaney stipulates a zero tolerance for carcinogenic residues in all food products. The threat to compromise that standard was an early signal that the Clinton administration would follow the Bush doctrine of negligible risk. If that wasn't enough of a hint that Browner had bought into third-wave environmentalism, an interview in *Fortune* magazine settled the matter. "We need to create incentives for plant managers in companies all across the country to look for ways to get the most pollution control for the least amount of money," she told the magazine, "I need those guys working with me."[8] Risk assessment eventually joined unfunded mandates and takings to become the third prong in the "unholy Trinity" of backlash against environmental legislation and rulemaking. It was a devastating triad for a movement in decline. All three are at the heart of Newt Gingrich's "Contract with America."

Also early in the first year of the administration—and in spite of campaign promises to the contrary—Al Gore signed off on a test burn of Waste Technology Inc.'s massive garbage incinerator near a working-class neighborhood in Liverpool, Ohio. Anti-toxics activists, who had been fighting incinerators nationwide, saw his approval as a major betrayal. They quickly responded with a new slogan, "Al, read your book," and a mantra, "Where's Al?"[9] Washington enviros occasionally chanted the mantra, but they only mumbled over the incinerator.

The vice president's sudden distancing from environmental concerns frustrated Washington enviros so much that they took to writing Clinton to get Gore's attention. At a meeting in March 1993 Gore asked a group of environmental leaders why they had not come directly to him on a specific issue. Evidently Clinton had asked him about "this letter from your friends." When they explained that they had found him to be inaccessible, Gore became upset. At this point one of the enviros said: "You can't take us to the woodpile," and Gore adjourned the meeting.[10]

After repeatedly scolding George Bush for his inaction on global warming and promising on Earth Day 1993 to bring U.S. emissions back to 1990 levels before the year 2000, Clinton announced new air quality regulations. He portrayed them as an effort to satisfy the terms of the Convention on Climate Change signed by more than 160 countries after the 1992 Earth Summit in Rio de Janeiro. Clinton's standards, however, were essentially toothless. They called for voluntary rather than mandatory compliance and failed to address the problem of controlling the projected increase in post-2000 emissions.

All but the most committed third-wave environmentalists saw the new regulations as another major betrayal. Even a consummate deal maker like Fred Krupp quipped that Clinton was "too eager to cut a deal." In New Hampshire, during the election campaign, Clinton had told a gathering of environmentalists: "I support an increase in corporate average fuel economy (CAFE) standards from the current 27.5 miles per gallon. . . . The 45 mpg. standards should be the goal of automakers and incorporated into national legislation." He repeated that sentiment to the League of Conservation Voters and printed it in his campaign literature. During his first year in office, however, Clinton not only allowed the auto lobby to block appointment of 14 separate candidates to the National Highway Traffic Safety Administration (which would enforce the CAFE standard), he refused to raise the standard above 27.5 mpg.

There were other divergences from the clean air standards supported by environmentalists. In 1994 Clinton allowed former NRDC lawyer Mary Nichols, who was appointed to oversee clean air matters at EPA, to cut a deal with Detroit car makers. The arrangement overrode the auto emission levels proposed by 13 eastern states that wanted to emulate California's strict clean air standards. Many economists and business development specialists believe that 14 states mandating such high standards would have created sufficient demand for electric cars to stimulate a booming new automobile industry in America. The administration handed enviros yet another clean air disappointment by accepting a utilities industry proposal. The plan provided tax incentives and pollution dispensation for planting trees, which, the utilities claimed, would become "sinks" for atmospheric carbon. "This is a plan that George Bush could have produced," complained Dan Becker of the Sierra Club.

When Interior Secretary Babbitt took office in 1993, he promised coal miners and their families to uphold and defend the Promise, as the Surface Mining Control and Reclamation Act of 1977 (SMCRA) is known in the coal fields. He would replace the lax and obstreperous administration of the U.S. Office of Surface Mining with a tough-minded regulator who would protect communities and Indian reservations from illegal strip mining. Babbitt appointed his friend Robert Uram, an industry lawyer who had spent much of his career defending coal companies that violated SMCRA. Babbitt and other members of his top staff are so heavily invested in coal that on several occasions they have had to recuse themselves from major decisions involving western coal mining.

Later Babbitt did take a tough stand against an 1872 mining law that virtually gives federal land to hard-rock miners. When a federal judge ordered him to sign an agreement to sell 1,800 acres to American Barrick Resources, a Canadian mining company, Babbitt, protesting that it was the "biggest hold-up in American history," nonetheless signed the sale agreement. It gave over $10 billion-worth of gold to Barrick.

As time wore on, the list of disappointments and Clinton administration betrayals grew.

• After Congressman Dan Rostenkowski, at the urging of the electrical utility industry, opposed Clinton's imaginative BTU tax, the president backed off. Since then he has taken great care not to antagonize Democratic friends in the oil and gas industry.

• Ignoring both environmentalists and the fishing industry of Washington, Oregon, and Idaho, the administration canceled a round of emergency spills at federal dams on the Columbia and Snake rivers designed to give salmon fingerlings a clear run to the sea. House Speaker Thomas Foley (D-WA) and other northwest pols persuaded the president that the releases would inflate the cost of hydroelectric power and drive heavy industry from the region.

• Pesticide activists were jolted again when Bruce Babbitt brokered a deal that allows Florida sugar farmers to continue polluting the Everglades with silt, phosphorus, and assorted toxic runoffs. He also surrendered to the sugar industry's demand that cleanup costs in the Everglades be shared by their industry and taxpayers, with taxpayers bearing the lion's share.[11] Political expediency and campaign finance evidently subsumed environmental concerns. Babbitt cut the deal with sugar grower Alonso Fanjul, chairman and CEO of Flo-Sun Corporation, a fundraiser and generous contributor to the Clinton campaign in Florida.

• After talking tough about grazing and mining reform, Babbitt and the president caved in to four western governors and half a dozen Democratic senators from Rocky Mountain states ("Rockycrats") opposed to increased fees. Even the Rockycrats, who had come to expect some increase in grazing fees, were surprised by the extent of Clinton's acquiescence. He abandoned altogether the tough grazing standards he and Babbitt had promised and placed stewardship of BLM grazing reserves in the hands of local citizens' committees. Later, when grazing fees became subject to a Senate filibuster, Clinton surrendered again. A few weeks later Babbitt fired the uncompromising and politically embarrassing Jim Baca from the top job at the Bureau of Land Management after Baca refused to accept a "promotion" that would have removed him from grazing and mining decisions.

• Although DuPont voluntarily agreed to cease production of ozone-depleting chlorofluorocarbons (CFCs), Clinton asked them to delay their action for a year. Executives from the big three domestic automakers had convinced him and EPA head Browner that the costly retrofit required to switch automobile air conditioners from CFCs to replacement HCFCs would cause a consumer revolt.

• In forming the President's Council on Sustainable Development, Clinton illuminated his own (as distinct from Al Gore's) environmental politics in three ways. He first asked the panel to recommend "a national sustainable-development action strategy that will foster economic vitality." Second, he appointed to the council executives from corporations on almost every environmental hit list in the country—including Dow, Georgia Pacific, DuPont, Ciba Geigy, Browning-Ferris, and Chevron. Third, he "balanced" them (one supposes) by NWF President Jay Hair, EDF CEO Fred Krupp, John Sawhill, head of the Nature Conservancy, Michelle Perrault, former president of the Sierra Club, and the NRDC's John Adams—some of America's most accommodating pro-corporate environmentalists. All but one of these environmental leaders (Perrault was the exception) supported the North American Free Trade Agreement (NAFTA) between the United States, Canada, and Mexico.

• Having secured an environmental side agreement to NAFTA with the assistance and support of conservative environmental leaders, Clinton attempted to barter away a vital environmental safeguard—the proposed banning of methyl bromide, an ozone-depleting pesticide—to win congressional votes for the treaty.

Nowhere did Clinton's strategy of obtaining reverse access to the environmental movement pay off better than in this aggressive campaign for support of NAFTA . When he came into office most national environmental organizations were either hesitant about or opposed to the terms Carla Hills had negotiated for the Bush administration. The agreement included no provisions to protect the environment or the environmental legislation of any of the signatories. Clinton sensed, probably correctly, that he couldn't win approval for the agreement in both houses without support from the environmental community. And he couldn't win that support without adding some minimal environmental provisions to the agreement.

Conservative enviros like the National Wildlife Federation's Jay Dee Hair had supported "fast-track" approval for NAFTA as a gesture of trust in Bush.[12] Nonetheless, he and most of the leadership remained neutral on the accord itself until environmental safeguards were clearly defined. The administration, however, needed more than that tacit approval. So Clinton sent Trade Representative Mickey Kantor and EPA Administrator Carol Browner back to the table with Mexico and Canada to negotiate an environmental side agreement, the North American Agreement on Environmental Cooperation. NAAEC created a multilateral commission to monitor each country's environmental enforcement and committed $4 billion to clean up the toxic trough left along the U.S.–Mexican border from a 30-year experiment with regional free trade. NAAEC, which was widely supported by polluting corporations and most of the mainstream environmental leadership, virtually prohibits public participation in the process that determines whether trade affects pollution control, resource conservation, or sustainable agricultural production in any of the three signatories.

Kantor brought Jay Hair, Justin Ward from NRDC, and a few other key environmental leaders into the negotiations. Hair joined the Kantor trade investment advisory committee, traveled south to meet with representatives of Mexican nongovernmental organizations and government officials, and returned home an ardent supporter and active lobbyist for the treaty. His most aggressive lobbying, however, was reserved for uncommitted environmentalists—particularly Carl Pope and Michael McCloskey at the Sierra Club and Jane Perkins at Friends of the Earth. Other leaders already committed to NAFTA, including NRDC's John Adams and the Nature Conservancy's John Sawhill, also joined trade advisory committees, where they "negotiated" with corporate executives and the American delegation on behalf of the environment.

Environmental appointees, some of whom had led the fight against fast-track legislation during the Bush administration, were invited to work in the NAFTA "War Room" with strategist Bill Daley. It was another smart move on Clinton's part. No one understood the opposition or could anticipate its arguments better than someone who had once been part of it. The administration also targeted specific environmental groups by granting small favors. National Audubon, for example, was assured that the agreement would protect migratory birds.

One by one, former NAFTA opponents and skeptics became enthusiastic supporters, and said so publicly. Eventually, leaders of Conservation International, the Environmental Defense Fund, the National Audubon Society, the National Wildlife Federation, the Natural Resource Defense Council, and the World Wildlife Fund formed the Environmental Coalition for NAFTA and threw their unequivocal support to the agreement.[13]

In the spring of 1993, with debate over the agreement heating up in Congress and the vote too close to call, members of the coalition appeared publicly with Vice President Gore to express support for the treaty. Even more significantly, they dropped their demand for creation of a trilateral North American Commission on the Environment empowered to subpoena documents and issue sanctions against violators of environmental standards spelled out in the side agreement. Once again the ancient Roman tactic of inviting one's critics in for a chat had paid off for Clinton and Gore.

As what appeared to be a very close vote approached, Gore and Kantor exaggerated pro-treaty forces by claiming support of almost all American environmentalists: "Groups representing 80 percent of national group membership have endorsed NAFTA," they proclaimed in a joint release. It was, of course, an outlandish boast. The total membership represented by the Environmental Coalition, 4.8 million, was outnumbered almost three to one by the nearly 300 regional and national environmental organizations (including 12 local chapters of the Audubon Society) that formed the Citizens Trade Campaign—which openly opposed the treaty.

As the vote drew closer the discord between pro- and anti-NAFTA enviros became rancorous. It was perhaps the nastiest internecine squabble in the movement's hundred-year history. NWF's Jay Hair, who became an active lobbyist for the agreement, openly accused those who opposed it of "putting their projectionist polemics ahead of concern for the environment. They haven't seen the documents," he said, "they don't know the facts and are feeding off misinformation."[14] When Defenders of Wildlife chief Roger Schlicheisen suggested that NAFTA might not sufficiently protect marine mammals, Hair berated him for raising "phony environmental arguments."

Hair also wrote caustic letters to hesitant environmental leaders like the Sierra Club's Michael McCloskey: "I will not mince words. We have been friends for many years. I value that friendship, and I admire your commitment as an environmental advocate. . . . [But] I regret that the Sierra Club has associated itself with organizations that do not value credibility as highly as I know you do and are willing to misrepresent the truth in order to prevail. But, of course, it is your decision to choose with which organizations you form alliances." To Jane Perkins at Friends of the Earth he wrote patronizingly that it was "important to question those who use obfuscation, dissembling, omission, overstatement, and occasional fabrication to advance their cause. I find it particularly distressing when it occurs among the environmental community, which ordinarily and with great pride occupies the high ground."[15]

Controversy over NAFTA reached deep inside the mainstream organizations. High-level staffers at EDF, for example, had grave doubts about the agreement but could not convince their boss Fred Krupp, who spoke for them all in endorsing the agreement. NRDC's field offices divided bitterly over NAFTA, even after the side agreement was attached. NRDC chief John Adams nonetheless asserted after the treaty was signed that Gore's 80 percent claim of environmental support was "probably close" and that a similar majority of his own staff supported NAFTA. "I'm sure of it. I was in all the meetings," he insisted, "and we had thousands of hours of meetings over this issue."[16]

Adams' interpretation of the final outcome reveals the loyalty to the administration some mainstreamers maintained long after signs of betrayal were evident on other issues. "We were one of the two big prongs the administration had to fight," he asserts. "The other was labor. We broke the back of the environmental opposition to NAFTA. After we established our position Clinton only had labor to fight. We did him a big favor."[17]

Support of NAFTA by mainstream leaders seems shortsighted, given their level of education and global sophistication. Carla Hills' assertion that prosperous, developed countries are more likely to protect their environments than poorer nations may be correct. Kantor's side agreement did provide some enforcement of standards and protection of hard-won environmental legislation. Nonetheless, it should be clear to any environmental thinker that free trade can only lead to the globalization of massive, consumer-based economies that are, in the long run, whatever the legislated safeguards, ecologically destructive. But mainstream environmental officials evidently don't think a lot about the distant future. Like the corporations they have come to resemble, they tend to be occupied with the day-to-day imperatives of strategy, competition, and survival. Because they have so deeply immersed themselves in the governmental domain they feel compelled to enter every political fray, no matter how tenuous its environmental connection.

The next test of trade policy for the nationals came with GATT, the General Agreement on Trade and Tariffs. Most leaders who endorsed NAFTA had serious reservations about the GATT agreement, which they believed would seriously jeopardize state and federal consumer and environmental legislation—particularly food safety, toxic substances and waste-reduction laws—as well as the Marine Mammal Protection Act, CAFE standards, the Magnuson Act (regulating fishing), and the Nuclear Non-Proliferation Treaty. Clinton, however, proceeded as if he didn't need support from environmentalists. By late 1994 the newly elected Republican majority seems to have endorsed the agreement. At the close of the 103rd Congress, GATT passed both houses, with barely a whimper from the environmental lobby.

When candidate Clinton promised to reform the historically corrupt and environmentally destructive U.S. Forest Service, and when as president he held the Forest Summit to create a plan to protect forest ecosystems "in a manner consistent with environmental laws," the entire environmental movement believed him. After hosting the summit, held during his first year in office, he was considering eight options to protect the remaining ancient forests, the spotted owl, and timber industry jobs. With an assist from Secretary Babbitt, Clinton selected a ninth option, which allowed continued, albeit reduced harvesting of publicly deeded old growth timber throughout the northwest, even in spotted owl habitat.

Despite the fact that half the remaining ancient forests in the United States could eventually be harvested under the plan, that the other half was not protected by being designated inviolate reserves, and that scores of plant and animal species were endangered by it, most mainstream environmental organizations eventually supported "Option 9." "A fair and reasonable compromise," Carl Pope, executive director of the Sierra Club, told *USA Today*, while National Audubon's Brock Evans called it "a shaky victory." Some leaders even declared it a minor triumph for the ancient forest and the environmental movement. National Wildlife Federation lawyer Tom France opined in the *Daily Missoulian* that "vigorous timber cutting is not incompatible with ecosystem management." Even to grassroots forest activists inured to the mainstream nationals' acquiescence that seemed like extraordinary support—given the fact that Option 9 clearly violated the environmental standards of the National Forest Management Act and other federal legislation lobbied into law by the Sierra Club, National Audubon, and the National Wildlife Federation.

Grassroots activists were particularly horrified by Tom France's statement, and by a letter Brock Evans wrote from Washington. In it he said that two Oregon-based activists who opposed the option were unable to make rational judgments about the situation because they were "3,000 miles away from what is really going on." It was another example of the Beltway hubris that grassroots environmentalists find so objectionable. "Where is 'what is really going on,' " asks Tim Hermack one of the Eugene, Oregon, forest activists Evans was talking about, "Washington

D.C. or Washington state? Brock Evans can't hear the chain saws in the Breitenbush Cascades from his air-conditioned office on Pennsylvania Avenue. How can *he* know what's really going on?"[18]

After the Forest Summit and implementation of Option 9, a coalition of national and local environmental organizations filed for an injunction against all timber sales in spotted owl habitat. In late September 1993, the administration threatened to override environmental laws, including the Endangered Species Act, unless the plaintiffs allowed resumption of logging, including clearcutting, on selective parcels of the national forests—including some in spotted owl habitat. The plaintiffs responded by agreeing to release 83 million board feet of timber for cutting if the administration would pledge to oppose restrictions on judicial review of federal timber sales. The administration countered by insisting that the plaintiffs would have to release 100 million board feet or they would ask Congress to exempt the timber industry from environmental laws.

Jeffrey St. Clair, an Oregon-based forest activist working with Save the West, said this about the Clinton administration: "Dealing with Clinton ain't like dealing with Reagan or Bush. In fact 'Team Clinton' is much more dangerous because they know our movement inside out. Bush didn't ask us for any favors, and none were given, but when St. Albert and Brother Bruce [Babbitt] came calling, attention was paid immediately."[19]According to St. Clair the result was defeat.

Not many mainstream environmentalists see it that way of course. Although at first most opposed Option 9 and subsequent timber sales, 12 national organizations eventually sided with Clinton on both issues. Babbitt's response is revealing. "We have secured from the environmental movement a commitment to work with us and not against us in designing a stable defensible timber sales program. They have committed to continue and expand this process." Though oft touted as a "win-win" solution, Option 9 is in fact a lose-lose-lose-win situation. The loggers lost, the enviros lost, and the forests lost. Only the president won by looking to the media and the rest of the country like a brilliant mediator.

Jeff St. Clair struggles to explain how environmentalists could turn "from defenders of ancient forests to designers and defenders of timber sales in ancient forests. Despite a reputation for confrontation," he asserts, "most environmentalists are romantic idealists by nature. They desperately want to believe that 'enlightened' people are fundamentally good and inclined to do the right thing. The Clinton gang, they believe, are environmentalists at heart; many in the administration actually worked for environmental groups. So if anyone was worth taking a chance on it was this administration, right?" The motive of the nationals, which, St. Clair believes sold out the regionals and the ancient forests, was to build "a relationship of power and control, not good will, cooperation, and trust."[20]

The tendencies St. Clair sees in mainstream leaders were much in evidence during Clinton's first year in office, as Washington enviros focused almost exclusively on the good news. His first day in office Clinton abolished Dan Quayle's Competitiveness Council and, shortly thereafter, signed the Biodiversity Treaty that Bush had refused to sign at the Earth Summit in Rio. He reversed the previous administration's population policies, which refused to fund international family planning organizations that countenanced abortion. He held the ancient-forest summit he had promised, increased funding for endangered species and renewable energy, ordered the government to purchase postconsumer recycled paper, and signed an executive order mandating government-wide use of civil rights laws to advance environmental equity in poor communities and requiring all federal agencies to "make environmental justice a part of what they do." Moreover, the Office of Management and Budget no longer attempts to limit environmental regulation with cost-benefit guidelines. Neither of Clinton's two predecessors would have done any of these things.

On the other hand, had Reagan or Bush performed even one or two of Clinton's manifold misdeeds, national environmental leaders would have called press conferences, roundly criticized the administration, and flooded the mail with urgent appeals for new members and money. Two years into their term, both Clinton and Gore were still being treated with relative equanimity by Washington greens. The League of Conservation Voters (LCV), which annually grades the president and Congress on environmental performance, awarded Clinton an overall passing grade of C+ and delivered a polite rebuke for "not living up to his potential." "This is a president who gets it," extolled LCV President Jim Maddy upon releasing his report card. Later, however, Maddy admonished Clinton for failing to use the White House as a bully pulpit for the environment. "He talks about the environment only on Earth Day," Maddy complained. "Between Earth Day 1993 and Earth Day 1994 he said nothing."[21] And in the Oval Office Clinton defers repeatedly to the National Economic Council and, as Maddy ruefully admitted in his assessment of the administration, "[together] they lapse into old ways of thinking. They appear to believe that environmental protection is an economic luxury, a lifestyle luxury, a political luxury for the President, a luxury that comes somehow at the expense of economic growth."

As shifty and unpredictable as he is, Bill Clinton nevertheless provides the mainstream environmental movement a functional proving ground for its heaviest weapon—the lobby. If the environmental lobby, with its 25 years of experience in Washington could not affect a Democratic Congress and a rhetorically friendly Democratic administration, it may never be effective again. By all indications, this is the case. Every piece of existing environmental legislation brought to the Hill for revision is weakened in the process, with the apparent approval of the White House. With the weak exception of the Desert Protection Act, not a single piece of significant legislation was signed into law during the first two years of the Clinton administration. Smart observers began to conclude that it was safer not to tinker with the Clean Water Act, the Superfund law, the Safe Drinking Water Act, the Endangered Species Act, or Wetlands Protection. Even before the 1994 elections, when lobbyists could count the environmentally friendly congressional representatives on the fingers of one hand, the president was apparently enamored with voluntary compliance and Al Gore was very hard to find.

Between 1969 and 1990 the Washington-based environmental lobby grew from 2 to almost 100 registered representatives.[22] By the early 1980s Senator Proxmire had proclaimed it "the most effective lobby in Washington." Pound for pound, Proxmire may be right. But dollar for dollar is what counts in Washington. National environmental organizations still invest millions of their members' contributions in the environmental lobby every year. It's a large portion of their combined budgets and seems like a lot of money until one considers the investment of industry's counter-lobby, which is measured in billions. Every industrial lobbyist who heads for the Hill to fight an environmental rule or regulation is backed up by PAC funds, many of which are larger than the entire environmental lobbying budget. In Washington a lobby without a PAC behind it is like a home buyer with a part-time job. "You're just not taken as seriously," according to veteran Sierra Club lobbyist Dan Becker, "even when you have the facts on your side."

The environment has PACs too (see Chapter 4), but they are minuscule. In 1992 pro-environmental organizations contributed about $1.3 million to congressional candidates. The same year the energy and natural resources industries alone gave almost $22 million. Add to that an undeterminable amount donated by other industries contesting environmental regulation and the environmental effort seems pathetic.[23] "The most I can offer a congressional representative is $10,000—$5,000 for the primary and $5,000 for the general election," says Becker. "That seems like a lot until you consider that the guy coming in after me from the Chemical Manufacturers Association has a hundred companies behind him, each with their own PAC that can offer $10,000 a piece. I'm out-gunned."[24]

"Lobbyists are suffering from the delusion that they have clout because they have the name of a national organization behind them," says Stewart Brandborg, former head of the Wilderness Society. James Dougherty, a vice president of Defenders of Wildlife and a member of the Sierra Club board of directors, has grave doubts about the future of the lobby.

> Legislative advocacy at the federal level—traditionally the environmental movement's mainstay—now requires an ever increasing amount of hard work to make significant progress. . . . [Compare] the relatively easy campaign to enact the Superfund law in 1980 with the 11-year struggle to amend the Clean Air Act in 1990. In the 1970s we looked to federal agencies for progressive stewardship of this nation's resources, and with them formed partnerships such as the one that won enactment of the 1980 Alaska Lands Act. Since then our energies have gone largely to overcoming bureaucratic intransigence.

Dougherty announced that his organization was beginning to explore "new methods," including direct action against corporations and grassroots organizing.[25]

James Monteith, former executive director of the Oregon Natural Resources Council, believes that the environmental movement's "power on Capitol Hill fifteen years ago to leverage votes on tough issues was twice what it is today, even though the environmental community now has four times the money and gobs of 'access.' . . . Our obsession with access and inside dealing has reached a new high with this 'friendly' administration."[26]

The environmental lobby has also failed to accept a critical fact about lobbying Congress: appropriations are as important, if not more important, than authorizations. An authorization without money behind it, like an unfunded mandate, is a vulnerable if not useless piece of legislation. It is easier, and from a promotional standpoint more advantageous, to lobby for an authorization than for an appropriation. "We persuaded Subcommittee X to amend the Clean Water Act. . . . " makes much sexier direct mail copy than, "our efforts to obtain an appropriation for amendment X of the Clean Water Act enforcement paid off." Furthermore, appropriations committees are generally tougher nuts to crack than authorization committees. The subcommittee that funds EPA, for example, also funds the Department of Housing and Urban Development, the National Science Foundation, and NASA. That's a lot of heavy political competition for environmental lobbyists, who would just as soon move on to the next authorization as compete for funds with space stations and low-income housing.

Some environmentalists believe that the mainstream lobby is focusing on the wrong issues, while others claim the focus is too narrow. "If the mainstream movement persists in lobbying Washington," asserts Rainforest Action Network Director Randy Hayes, "they should lobby as hard for campaign reform as they do for environmental issues. Since the movement will never be able to match industry's war chests, the only way to level the playing field is to take their money away, and the only way to do that is through campaign finance reform."[27] Most environmental leaders and their lobbyists accede to Hayes' argument but few are doing anything about it. Meanwhile the movement is losing ground on the issues it does focus on—toxics, wilderness preservation, takings, risk assessment, unfunded mandates, clean air, and clean water. Yet the rationale for lobbying is not being questioned; it has become so central to mainstream environmental strategy that shifting resources away from it would seem like a retreat from battle.

It is clear now that—with or without a lobby—the environment is not going to receive anything like the attention Bill Clinton gave to his (losing) health care plan. Only a few seconds (117 words to be precise) of his hour long 1994 State of the Union address were addressed to the environment. It also seems clear that Clinton knows the nature and relative value of the environmental vote. Basically it's his to lose.

The Green Vote: Real or Ephemeral?

A well-organized minority can often defeat an unorganized majority.
Richard Viguery, Conservative Fundraiser

The environmental movement has become the political equivalent of the children's crusade, so politicians don't really have to pay attention to it.
Donald Snow, Northern Lights

A central assumption of the mainstream movement's lobbying strategy has always been that there is an "environmental vote"—a block of voters who allegedly cast their ballot for the environment above all other issues. Implicit with every visit from a Sierra Club or NWF lobbyist is the threat that a few thousand voters out in the district might vote for a challenger if the representative lets them down on this important bill. The League of Conservation Voters, formed in 1970, bases its operations in Washington and in key legislative states on that central premise.

It is not an invalid assumption. Environmental sentiment does run high in the United States. Despite the numbers claimed by Wise Use leaders and other antagonists, a May 1992 Roper poll found that 80 percent of Americans were concerned about environmental issues. An Environmental Opinion Study (EOS) poll taken in 1994 for the League of Conservation Voters (LCV) indicates that about 87 percent of voters still consider themselves environmentalists. This figure is down a bit from the numbers reported during the late Reagan years but is much higher than those from similar polls taken in the early 1970s. It is still an impressive number to any prospective electoral candidate—federal, state, or local—attempting to fathom the significance of the green vote. Almost 70 percent of voters also see a need for additional environmental regulation; and even in allegedly anti-environmental states like New Mexico and Montana local and regional polls show support for endangered species, wetland protections, and grazing reforms. On the other hand, the polls taken on broad national issues find that fewer than 5 percent of Americans rank the environment as one of the nation's most pressing problems, and very few actually identify themselves with the environmental movement.

In 1992, the George Bush campaign misread those polls and ignored the latent environmental concerns expressed in earlier surveys. Bush lost several western states by underestimating public support for the environment (or overestimating the strength of the Wise Use movement). That assumption certainly lost him the state of Washington. In the small town of Colville, Bush told a rally of lumbermen and mill workers that if Clinton and "Ozone Man" were elected America "would be up to its necks in spotted owls." The rally cheered him, but the sentiment bombed, not only in the state, but also in the surrounding county. Both went for Clinton.

EOS polls show that an equal number of Democrats and Republicans call themselves environmentalists but that Democrats vote the environment far more fervently and frequently than Republicans do. Of Democrats, 53 percent say that a candidate's endorsement by an environmental organization might affect their voting decision; only 40 percent of Republicans say they would be influenced. Moreover, the jobs-versus-environment tradeoff no longer seems to be an issue with voters. Fifty-five percent of those polled say that when the conflict arises they would side with protecting the environment; only 23 percent would side with jobs. The remaining 23 percent refused to make an abstract choice.

Yet it is worth questioning whether the green vote is as powerful and significant as the polls seem to suggest, and whether political candidates, incumbents and challengers, pay it much heed. Do those millions who call themselves environmentalists really vote their environmentalism, or does it actually have a lower priority for them? Focus groups and polls that probe more deeply seem to suggest that while Democratic women and liberal to moderate Republican voters are likely to act on their environmental concerns, only about half of all voters say that the environment actually affects which candidates they vote for. In short, most voters seem to have higher and often pre-emptive priorities.

In 1992 an election-night survey performed by EOS revealed, to no one's great surprise, that "the economy was the driving force behind most voters' presidential choice." Sixty-seven percent said it was the most important factor; 39 percent gave priority to education; while 29 percent said the environment was their top voting priority (42 percent of Clinton-Gore voters, 20 percent of Bush-Quayle voters, and 19 percent of Perot voters). That response substantiated an earlier EOS survey that found that about one-quarter of the American public "claim to have used the environment as a litmus test in their voting," while only 15 percent consider the environment as "one of the most important problems facing the country."[28]

Environment first-and-foremost voters "tended to be Democrats, women, especially Democratic women, voters under 30 years of age, and women without a college education." However, most voters in the 1992 national election (73 percent) said they did not consider the environment when they were deciding which candidates to support for various federal, state, and local offices. Only in the state of Washington did more than 30 percent think about environmental issues when choosing between gubernatorial candidates. Nonetheless, all state samples showed voters rejecting "George Bush's notion that high environmental standards will cost jobs, raise prices, and hurt the economy."

A slight majority of voters (51 percent) complained that the environment was not discussed enough during the campaign, and 10 percent said it was discussed too much. Even though only 29 percent made environment the most important factor in their presidential choice, a full 70 percent said that the environment should be either a top priority (20 percent) or a very important (50 percent) issue for the new administration. Figures were considerably higher for voters under 30. Another heartening sign for the future is an earlier EOS finding that Americans rank the environment right behind crime and drugs as "the greatest threat to future generations." Again, the responses indicated that the environment is a nonpartisan issue, with 76 percent of Democrats, 61 percent of Republicans, and 70 percent of independents saying that it should be an important priority for the president elected in 1992.

But is the president listening? If his first two years in office are any indication he either doesn't believe there is an environmental vote, is confident he can win it with a few trophy decisions during re-election year, or thinks that as a Democrat he can take environmentalist voters for granted. Congressional Democrats, including formerly dedicated environmentalists like Bruce Vento (D-MN) and George Miller (D-CA), also seem willing to ignore the environmental vote. Perhaps they fear it will hurt the party, which is, naturally, for most politicians a higher priority than the environment.

If the economy continues to improve it is possible that the American voter will give the environment higher priority. In that event if the Republicans run a moderate candidate with a decent environmental voting record (60 or better on the League of Conservation Voters scoring), the environment could become, for the first time in American history, a determining issue in a presidential campaign.

Power Breakfast

I genuinely fear that our experiment in democracy may be doomed.
Carl Pope, Sierra *Magazine, October 1994*

On Tuesday, February 3, 1994, Fred Krupp hosted a meeting of the Green Group at the offices of the Environmental Defense Fund in New York. Because the group's meetings are off-the-record, no minutes are taken and participants, mostly organization CEOs, are expected to be judicious about discussing the agenda with outsiders. Somehow, however, word reached the White House that the mainstream environmental leaders who attended that Tuesday meeting were losing patience with the Clinton administration. A few days later a group composed of the CEOs and chief lobbyists of the mainstream nationals were invited to breakfast at the White House with Al Gore, Carol Browner, Bruce Babbitt, and Tim Wirth.[29] Most people who gave me their views of the meeting were not willing to be identified. "Not much was accomplished," was the most common remark. Yet more than one participant also allowed that "it could be the beginning of something positive."

After coffee, sweet rolls, and small talk, Carol Browner sounded a call to arms on environmental issues. She said that both the movement and the administration have been taking a beating from industry and the Wise Use people. "We need to work together," she pleaded. No one disagreed. Al Gore had to leave breakfast early because of a death in his family. But before departing he mused, "There must be a nexus between environmental issues and health issues. If we could only find it."

Interpretations of his remark varied. Some participants believed the vice president was simply trying to hitch the environment to health care reform, which at the time was consuming more administrative energy than any single issue since the war in Vietnam. Less generous observers were shocked that Gore seemed unable to see the very obvious connection between health and environment.

"There was a general lamentation that we needed to get back on the balls of our feet on environmental issues," recalled Bill Roberts, who represented EDF in Fred Krupp's stead. "Get on the offensive, get aggressive, and be proactive, get the environment back on the radar screen, get the attention of Congress and the administration. That led to a discussion about the White House being consumed with health care reform." So Gore's remark, according to Roberts, simply said: "That's where everyone's head is—link environmental issues with health care reform. Every gun they have is working on that issue."

The same evening Al and Tipper Gore held a reception at the vice presidential mansion and invited all the meeting's participants and their spouses. Gore said that he had called the meeting that morning so that people would not arrive at his house, "as they so often do, with lists and agendas in their pockets." Many toasts were raised that night. Many commitments were made to continue working together, And the vice president suggested making the environmental breakfast a regular affair.

The day after the breakfast a few of those who attended met with Congressmen Henry Waxman and George Miller, both Democrats from California and two of the most aggressive environmental legislators in federal government. "Doesn't anyone notice that we are getting our asses kicked?" asked Miller, reciting a depressing list of defeats the environment had suffered during the past year. The two representatives were deeply concerned about the loss of congressional support for environmental issues. Miller described how Wise Use activists had tied up his office fax machines for two days on a single bill. Moreover, they both complained, they received no support from either the White House or the environmental movement. "They were begging for our help," according to Jane Perkins, then executive director of Friends of the Earth.

Perkins and her colleagues left the meeting, promising to place a conference call to all mainstream CEOs to jointly design a legislative strategy for the next Congress. It was a strange promise, given the fact that there had rarely been as much environmental legislation pending before Congress. Almost all of it was under heavy assault by forces far more potent than the environmental movement. By the end of that congressional session, however, it was evident that the best legislative strategy was probably to leave everything pretty much alone. Every existing act scheduled for reauthorization, particularly Superfund and the Endangered Species Act, had been targeted by environmental antagonists determined to weaken it. With minimal support for strengthening amendments or new regulations, it seemed foolhardy to push for reauthorizations.

Earth Day 25

On April 22, 1995, the unofficial publication date of this book, America will celebrate its 25th Earth Day. Preparations have been underway for at least two years. Organizers of small and large celebrations hope to restore some of the luster the environmental movement has lost since Earth Day 1990. Every major city will host dozens, if not scores of events, and almost every school and university will hold a teach-in. Politicians and business leaders from coast to coast will wrap themselves in green rectitude. Priests and rabbis will pray for the planet, Buddhists will meditate, trees will be hugged, and New Age pagans will converge with The Force. Even the Wise Use movement threatens to acknowledge the day in some way.

Since Earth Day 1, April 22, 1970, interest groups of all kinds have attempted to exploit Earth Day to their own advantage. But no group's participation has stirred more controversy than that of the business community, whose role as sponsor and participant has grown exponentially. It would be hard to imagine a more effective way to put a green spin on a corporate image than by sponsoring an Earth Day event. By all indications, sponsorship is for sale.

Bruce Anderson, an environmental business consultant and architect who was active in Earth Day 1990, was the first to move into the vacuum when Denis Hayes, the prime mover of Earth Days 1970 and 1990, retired from the job to chair Green Seal (an eco-labeling service) and then to run the Bullitt Foundation in Seattle. Preparations for the 1995 celebration began in 1991 when Anderson, Claes Nobel, and Gaylord Nelson formed Earth Day USA. Anderson's opinions are in sharp contrast to the sentiments expressed on Earth Day 1970. He believes that "we are all to blame, every one of us," a view that provides little basis for excluding anyone from the celebration. And indeed no one has been denied a sponsorship role. For $20,000 a company could become an official Earth Day booster and negotiate for permission to use the official Earth Day logo. For $7,000 almost anyone could also buy sponsorship of a newsletter sent to about 4,000 grassroots organizers. "If a business says they want to improve their environmental record, it's not up to Earth Day USA to be the judge and jury of their past behavior," said Anderson.

Gaylord Nelson also welcomed corporate sponsorship for Earth Day XXV and for the group planning it. "We're not going to have a sustainable society unless we have all interest groups on board," he told a public relations newsletter in early 1994, "I'm glad to see corporations joining in. If they try to co-opt Earth Day, they'll just help spread environmental propaganda. I'm not at all worried about greenwashing." And about the fact that Earth Day USA hired Dorf and Stanton, a PR firm that also represents Ciba Geigy, Chase Manhattan Bank, Ford Motor Company, Hydro-Quebec, Monsanto, Pfizer, Purina Mills, and many other corporate clients, Nelson seemed unperturbed: "Shouldn't we let the sinners repent?" But are they really repenting? Aren't they just greenwashing? "Greenwashing isn't even on the horizon as an environmental issue," Nelson retorted. "Besides, if people weren't running around claiming to be green we would have failed the course. If some of them are insincere about it that's all right. The market place will take care of the cheaters. They will be exposed by Earth Day 2000."[30]

Like Nelson, Anderson welcomed all comers to Earth Day USA until August of 1994, when dissent among the board members caused the organization to dissolve. "They were uncivil, rude, and irrational in their contempt for Bruce Anderson," says Nelson of two board members he declined to identify.[31]

With the national celebration in limbo and only a few months remaining, Denis Hayes was invited back to settle differences. He agreed to run Earth

Day USA, with the proviso that all board members tender their resignation. "Unfortunately, everyone did," lamented Hayes, who was left with an empty shell and a lot of potentially embarrassing stories about the sale of corporate sponsorships.[32] He quickly recruited Wally McGuire to resurrect and run Earth Day USA. McGuire, former director of the Environmental Policy Center in San Francisco and field operations director for Earth Day 1990, subsequently became a large events coordinator. He is "perhaps the best events coordinator in the United States," said Hayes of McGuire, who choreographed the Pope's 1987 visit and the Olympic torch marathon.

When questioned about whether Earth Day USA would sell corporate sponsorships, Hayes acceded that "it would be impossible in 1995 to celebrate Earth Day without corporate support." But Hayes said that unlike Earth Day XXV, Earth Day USA would not accept money from just any corporation. He would work with the socially responsible investment community to screen bad companies from the sponsorship list, using "relatively straightforward" criteria. "We are alert to companies whose involvement might cause both them and us to be viewed skeptically," Hayes told the *Green Business Newsletter* in September 1994. "However," he added, "participating companies would not be limited to Ben and Jerry's and Patagonia." A few days after saying that, Hayes resigned from Earth Day USA, which promptly folded.

Meanwhile Bruce Anderson kept moving. Together with Gaylord Nelson he formed Earth Link/Earth Day XXV. With a $25-million budget they hope to present "the seminal celebration for 1995"—an all-day extravaganza on the Capitol Mall in Washington featuring speeches by the president and vice president. Like the organizers of Earth Day 1980, Anderson and Nelson were reportedly promised $3.7 million in seed money from the White House. Within days of starting up, Earth Day XXV had raised $6.5 million in corporate contributions. With it they hired a "sponsorship marketing agency," World Premiere Marketing, to handle "corporate packaging and sales." "It's so crucial that corporations put money into Earth Day," Anderson told the *Green Business Newsletter*. "It takes tens of millions of dollars to do a respectable Earth Day."

The unfortunate thing about Earth Day XXV, if indeed it occurs, is that the big event on the Mall will almost certainly outshine the truly meaningful events of the day—the thousands of small celebrations, demonstrations, and teach-ins taking place around the country on the same day. The public perception will be that Bill Clinton is indeed an "environmental president" and that his administration is totally committed to protecting the environment from those few corporations that haven't yet greened up their act.

8

The Fourth Wave

Greens are openly contemptuous of politicos, politicos see greens as immature publicity seekers; grassroot activists think the politicos are a little too fancy and the greens are a little too far out; globals think the grassroot activists are a little too parochial and grassroot activists think globals are a little too abstract.

Walter Truett Anderson, Pacific News Service

Those of us committed to ecology as a subversive science, one which offers the possibility of an alternative, steady-state society for the twenty-first century, must continue to establish or revitalize locally active community environmental organizations.

Gilbert La Freniere

We cannot solve the problems we have created with the same thinking that created them.

Albert Einstein

Academicians who study the life cycles of social movements are often baffled by the endurance of the American environmental movement. For many it has long outlasted its life expectancy. Riley Dunlap and Angela Mertig of Washington State University describe the end stage of a typical movement:

In the process of achieving success a movement typically loses momentum. Its organizations evolve into formalized interest groups staffed by activists-turned-bureaucrats, many of its leaders are coopted by government to staff new

> agencies, or simply tire of battle, and support dwindles as the media turn to newer issues and the public assumes the problematic conditions are being taken care of by government. Efforts to revitalize the movement and avoid stagnation and co-optation may lead to rancorous in-fighting and fragmentation, with "die-hard activists" disavowing those coopted by government or seduced into working "within the system." . . . Well-intentioned agencies begin to fail, typically because they are captured by the very interests they were designed to regulate. These trends may result in the demise of the movement, as it disappears with little if any improvement in the problematic conditions that generated it.[1]

Has the American environmental movement escaped such a fate? It certainly suffers from many of the characteristics Dunlap and Mertig describe. Yet is has outlasted other American movements, partly because of the anti-environmental policies of the Reagan administration, which enabled sclerotic organizations to expand membership and hold on for another 10 years, but also because of the endurance of the American environmental imagination. As new degradations became evident and different aspects of the biosphere came under assault—some threatening the entire planet—new surges of activism rose to revitalize the movement. Revitalization seems to have eluded the mainstream sector of the movement, however. Today it is suffering a chronic loss of zeal, membership, and imagination.

The third wave of American environmentalism described in Chapter 5 is currently ebbing from the shore; in every respect it has been smaller than the previous two waves—and certainly less consequential. In a few years historians and sociologists may well describe it as a brief attempt by corporate America to capture the nation's environmental imagination. A more generous interpretation is offered by Francisco Sagasti and Michael Colby, advisors to the World Bank; they believe that "environmental protection is in a transitory phase made necessary by tensions inherent in the overly reductionistic, short-term view of frontier economics . . . which are characterized by the view that the environment consists of limitless resources."[2]

A distant observer of mainstream environmentalism and the activities of the federal government might conclude that the entire environmental effort was moribund. But as the third wave washes back into history a massive swell of new environmental passion is gathering force. Democratic in origin, populist in style, untrammeled by bureaucracy, and inspired by a host of new ideologies—the fourth wave should crest sometime early in the twenty-first century.

At present the fourth wave has no single defining quality beyond its enormous diversity of organizations, ideologies, and issues. It is part wilderness preservation, part toxic abatement, part ecological economics, part civil rights, part human rights, part secular, part religious, and parts of many ecologies. There are, however, some encouraging common characteristics among the parts. Anger, for example, and energy, a commitment to democratic processes, and its multivarious participants' urge to restimulate the environmental imagination and expand our concept of environment and environmentalism.

Most fourth-wave activists feel that the three previous waves of environmentalism have spent their relevance. They believe that what remains of the mainstream movement is simply too slow and accommodating for the task before it. A new nonviolent militancy is apparent. But the central sentiment that will define the next generation is, quite simply, a sense of justice, which until very recently has been almost completely absent from the American environmental imagination. Environmental equity, that safe phrase used by EPA officials anxious to avoid the *j*-word, will gain real meaning as rich and poor, white and nonwhite, mainstream and grassroot realize that all are living and toiling in the same environment. At that point environmentalism will begin to become a truly "social" movement

The fourth wave of American environmentalism will be *very* American. By all indications, the movement is already well on its way to becoming multiracial, multiethnic, multiclass, and multicultural. It also contains many traits that characterized the American Revolution—dogged determination, radical inquiry, a rebellion against economic hegemony, and a quest for civil authority at the grassroots.

Sociologists warn social activists that fragmentation is an inexorable consequence of expansion. But one scholar's fragmentation is another's diversity. It is true that diversity can be divisive, and there is certainly fractiousness among contemporary environmentalists. Deep ecologists deplore reformers, anti-toxic NIMBYs have little use for defenders of wildlife, and traditional conservationists decry eco-saboteurs. On the other hand, diversity has allowed the environmental imagination to expand into sectors and interest groups of society that had no use for environmentalism just a few years ago. Most academic disciplines now include a subdiscipline that focuses on some aspect of the environment. Almost every religious denomination in the country has a green preacher, priest, or rabbi, and most other clergy offer at least one sermon or homily on the subject every year (frequently on Earth Day). Civil rights leaders have placed environment on their agenda just as environmental visionaries have begun to explore a nexus with the human rights movement.

Although the next uprising of environmental activism will come from almost every sector of American society, it will be built principally from the combination of four national initiatives often regarded as completely separate social movements. The four sectors that will create the fourth wave are:

- Splinters from the mainstream movement—many of which have already joined one of the other three sectors
- The environmental ad hocracy—10,000 NIMBYs sharing tactics and coordinating strategy through such national organizations as the Citizens' Clearing House for Hazardous Wastes
- The people-of-color network for environmental justice
- The new conservation movement (most of which is currently involved with the preservation of forests on public land).

Splinters

Thank God for Dave Brower. He makes it easier for the rest of us to be reasonable.

Russell Train, World Wildlife Fund

Dave Foreman makes Dave Brower look like a raging moderate.

David Brower, Earth Island Institute

Throughout the history of reform environmentalism radicals like David Brower and David Foreman have collided with pragmatists in their own organizations. In their quest to perpetuate the organization that employed them, pragmatists developed a new meaning for the word *preservation*. Preserving institutions became a higher calling than preserving the environment. Tension between the two values resulted in splintering—the departure (or ejection) of people like Brower and Foreman. In 1969 the Sierra Club fired David Brower. Soon after, the Sierra Club Legal Defense Fund (SCLDF) led by Rick Sutherland declared itself independent of the club, which it found too slow and conservative for an effective litigative strategy. (SCLDF reluctantly kept its name because "Sierra Club" retains valuable identification and fundraising power.) The club itself may be on the verge of yet another split, which I will describe later in this chapter.

Dave Brower went on to form three separate organizations, Friends of the Earth (FOE), the League of Conservation Voters, and the Earth Island Institute—one of which (FOE) fired him again. Everything Brower has done has reanimated the movement as a whole. During the same years, scores of less well-known men and women left other mainstream organizations to work at the grassroots, where they felt more inspired, more effective, and more fulfilled.

Tiring of the self-preservationist tendency of *all* organizations, one splinter formed in 1980 had virtually no organizational structure or identity, only a name and a commitment to eschew all compromise "in defense of Mother Earth." Earth First! can be considered a bastard child of the Wilderness Society, where two of its founders worked. It was created by five "rednecks for wilderness" who decided—during a desert beer bust near Page, Arizona—to follow the example of Doc Sarvis, Seldom Seen Smith, and George Washington Hayduke, protagonists of the Edward Abbey novel *The Monkey Wrench Gang* (1976). Sarvis and his cohorts held, and acted upon, the belief that sabotage was the best, perhaps only way to save wilderness. "Ecotage" became the trademark of Earth First! and "Hayduke Lives" its war cry.

Leading the real life monkeywrenchers was David Foreman, the self-appointed "bad boy of the environmental movement" who is remembered by Washington enviros as an articulate, at times brash lobbyist for the Wilderness Society. While Foreman worked on the Hill, he and other frustrated enviros would sometimes discuss the need for "a more militant group that would be trotted out on occasion to take an extreme position and make the mainstream look more reasonable." In 1979 the U.S. Forest Service's 1979 Roadless Area and Review Evaluation (RARE

II) released 65 million of the Service's 80 million undeveloped acres to logging, recreation, and mineral development. Of the protected area, 7.5 million acres was rock and ice. To Foreman's dismay, the Sierra Club, Wilderness Society, and others declared victory. Foreman left the Wilderness Society, and Washington, in disgust.

Rather than applying for tax-exempt status and forming another nonprofit organization, Foreman and his fellow buckaroos camped in the Pinacate Desert and envisioned a new "tribe" of trained, loosely disciplined eco-saboteurs.[3] They would be a "nomadic action group" of modern-day monkeywrenchers—not "terrorists" as opponents would surely describe them, but saboteurs. "Sabotage is violence against inanimate objects: machinery and property," cautioned Edward Abbey. "Terrorism is violence against human beings. I am definitely opposed to terrorism, whether practiced by military and state—as it usually is—or by what we might call unlicensed individuals."[4]

"Do not harm human lives" became an early principle of Earth First!. Unfortunately it was not always followed, and Earth First suffered the consequences. Activists operating under the Earth First! banner created problems for their fellow travelers by embracing extreme ideologies. Some played on racist xenophobia, and others were just plain insensitive—like the pseudonymous essayist ("Miss Ann Thropy") who postulated in the May 1, 1987, *Earth First!* journal that AIDS was nature's way of protecting the Earth from excess population. "I take it as axiomatic that the only real hope for the continuation of diverse ecosystems on this planet is an enormous decline in population," read the column. "If the AIDS epidemic didn't exist, radical environmentalists would have to invent one." David Forman's remark, as starvation ravaged Ethiopia, that "the best thing would be to let nature seek its own balance" and let the people there just starve did not endear Earth First! to most people. In the same interview Foreman revealed an obstinate patriotism: "Letting the USA be an overflow valve for problems in Latin America is not solving a thing. It's just putting more pressure on the resources we have in the USA," he said.[5] "Send the Mexicans home with rifles," he exhorted on another occasion.

Tree spiking, the practice of driving large nails into trees with the intent of damaging chain saws, also turned out to be a self-destructive tactic for Earth First!. It was opposed by many Earth Firsters at the time, and is now regretted by many others. When a large industrial bandsaw in a northern California mill broke and lacerated a mill worker, the mill owner blamed a spike found in the log. It was the worst public relations crisis Earth First! would face until Foreman and four others were arrested in Arizona and tried for attempting to sabotage a nuclear power plant. They were convicted, and some are serving jail terms. Foreman, who has now distanced himself from Earth First!, is out on probation and active in building the new conservation movement.[6]

The splintering process has sustained a feisty, sometimes militant edge to the environmental movement. There have been many other splinters. Of note are the Native Forest Council started by Sierra Club dissident Tim Hermack in Oregon; Restore: The North Woods, the brainchild of former Wilderness Society employee Michael Kellett; the Sea Shepherd Society founded by former Greenpeace skipper Paul Watson; the Oregon Natural Resource Council, an authorized splinter of the Wilderness Society; and a host of lesser known groups with lyrical names like Heartwood, Greenfire, Headwaters, and Great Old Broads for Wilderness.

At the root of this splintering process is vision, something that large bureaucratic organizations in any movement have difficulty sustaining and acting upon. It was vision that drove Mike Kellett from the Wilderness Society. He left, he says, because "number one, I didn't believe there was enough of a grassroots movement in the region [New England] that involved all of the people who were out there, and concerned, but who don't know what to do to protect the environment. And number two, I felt that for the most part other groups in the Northeast were interested in focusing only on protecting what's there now, and not on restoring the ecological integrity of the region. I felt a group was needed to promote in a visionary manner the restoration of that integrity."[7]

Most, but probably not all mainstream organizations, will survive the splintering process and perform vital functions as part of the fourth wave—educating the middle class, litigating for nature, and holding the line against the industrial lobby. If they persist in their deferential ways, however, they will become even less effective and less relevant than they are now. Splintering from their ranks will continue, as their renegades join forces and network with grassroots troublemakers that the nationals have always been kept at arm's length. Together the new groups will form the fourth wave.

Forests Forever

At the heart of the new conservation movement is the battle to save America's ancient forests. Few struggles better illustrate the splintering of the mainstream movement and the emergence of fourth-wave environmentalism.

Responsibility for the nation's forests has always been a major item on the agenda of the traditional conservation movement. Yet in recent years mainstream wilderness and wildlife organizations like the Sierra Club and the Wilderness Society have left the good fight for the remaining 5 percent of America's ancient, old-growth stands—most of them scattered about on public land—to a host of local and regional organizations.[8] The battle fought by these groups illuminates the enormous perceptual differences between national and regional organizations and their conceptions of victory and defeat.

The main threat to ancient forests is the unfortunate, some would say corrupt policy of the U.S. Forest Service, which virtually gives trees to the timber industry and subsidizes the building of roads that loggers need to get the timber out. During the five years between 1988 and 1993, this policy cost taxpayers about $1.1 billion. In fiscal 1992 alone losses in 101 national forests exceeded $300 million. The practices the Forest Service allows or encourages on public land threaten wildlife habitats and the future of anadramous fish stock (a vital human and animal food supply and a $1.0-billion industry).

During the 1970s, when timber industry ties to key congressional leaders in Washington made it increasingly difficult to contest timber sales in old-growth forests, particularly in the Rockies and the northwest, regional and local conservation groups turned for support to the nationals who had offices in Washington. For years the nationals had boasted to their members, many of whom were also members of grassroots forest groups, that they were uniquely positioned to protect forested wilderness on western federal land. When called to join forces with grassroots efforts the nationals half-heartedly joined the fray. They also took great care not to offend congressional representatives with whom they had established trust, or who could be helpful with other items on their federal wish list. Not one single national, for example, has expressed opposition to clearcutting in wildlife refuges or logging in national forests. They are, however, all unequivocally opposed to clearcutting the ancient forests of Brazil, where 85 percent of the rainforest remains intact. But in America, where less than 5 percent of the ancient forest survives, they are silent. Nor, with the occasional exception of the Sierra Club Legal Defense Fund, have the mainstream organizations opposed the forestry practices employed on private lands, which are often far more environmentally destructive than those used on public lands.

In 1988, desperate at the accelerated loss of trees that had existed long before European discovery of the continent, the Ancient Forest Alliance was established to secure permanent legislated protection of old-growth forest ecosystems. Regional organizations soon realized that they had made one fatal mistake. They had allowed the National Wildlife Association, the National Audubon Society, the Wilderness Society, and the Sierra Club to join the alliance. Before long "The Big Four" had taken control of strategy, which soon became what long-time executive director of the Oregon Natural Resource Council James Monteith calls "a strategy of capitulation" to the timber industry and their representatives on the Hill. Congress, particularly the appropriations committees of both houses, became staging areas for repeated assaults on public forests. "Our reluctance to confront the forces of darkness," Monteith recalled, "resulted in unprecedented confidence and arrogance on the part of the congressional champions of industry. . . . The insanely high cutting levels of the 1980s and 1990s were set not by the Forest Service or the Bureau of Land Management, but by western politicians in the annual appropriations bills."[9]

As it became clear that the Washington-based nationals were refusing to hold the line on the harvesting of old-growth timber, dedicated forest activists began to ignore them and form small (sometimes one-person) groups. Today there are hundreds, perhaps a thousand such organizations across the country—over 175 in Oregon alone—each fighting to save a single forest or a single watershed from the chain saw. "It's the only way to do it," says Andy Mahler, founder of Heartwood and a central figure in the new conservation movement. "Legislation has to be constantly enforced and litigated. We can't possibly do that against the Reagan courts. So we defend specific places. It's a safe way to protect the forests because we know that no one in our movement is willing to give up someone else's forest."[10] They meet often, this rag tag movement of "tree huggers"; and when they aren't meeting they are networking through fax and e-mail. Although they often disagree on policy or tactics, most seem committed to the strategy of saving America's forests one timber sale at a time—reminiscent of Lois Gibbs' method of "plugging the toilet" one NIMBY at a time. It's tiresome and thankless work at times, but it may be the only way. As Stephanie Mills laments, there is "no time to await the dawn of that happy day when all of humanity sees the light. Salvaging these fragments of eco-systems . . . is desperate work. It's like rushing into a burning house to save the babies."[11]

Insurrection

They're all just like Queen Victoria, old, fat, and unimaginative.
Anonymous Sierra Club Dissident

The same regional/national friction that plagues the grassroots and the nationals exists to a lesser degree within some of the Green Group organizations themselves—notably the Sierra Club and National Audubon, which, although constitutionally more democratic than other organizations, frequently act without the advice or consent of members or regional chapters. The result is internal splintering that could lead, in both cases, to real organizational schisms.

In 1989 the Washington office of the Sierra Club helped Senator Mark Hatfield (R-OR) draft an amendment denying private citizens the right to sue against timber sales on the Oregon coast. Club officials moved without consulting their own grassroots members or other organizations and gave away far more than they needed to. Local Sierra Club activists, who had successfully protected huge stands of old-growth timber by challenging public sales to Georgia Pacific and other corporations, were horrified. So was the National Audubon Society, the National Wildlife Federation, and the Sierra Club Legal Defense Fund. All were ignored.

The club's leadership caved in again, according to local members, by countenancing the Sierra Accords, a provision allowing additional harvesting of old-growth timber in California. In 1990, after the close (52 to 48 percent) defeat of Proposition 130, the so-called Forests Forever referendum—the Sierra Club agreed to enter alternative dispute resolution negotiations proposed by Sierra Pacific Industries, the state's largest timber cutter. It was another classic case of the stronger party in a dispute choosing negotiations over litigation or other more forceful methods of resolving differences. The Sierra Club leadership should have known better.

Sierra Pacific's aim was to draft a new state forestry bill to be introduced in Sacramento and voted up or down, without interference or amendments from legislators. To the dismay of the forest activists (including many Sierra Club members) who had launched the Forests Forever campaign, the club agreed to a fast-track negotiated-settlement agreement in which it agreed to lobby other nationals, particularly National Audubon, NRDC, and the Wilderness Society. Grassroots organizations were never mentioned. In return, Sierra Pacific would attempt to bring timber associates like Louisiana Pacific, Simpson, and Georgia Pacific into the agreement. The accord drafted by the "two Sierras," known as the Sierra Accords, allows half the state's ancient forests to be cut every 25 years. Such a drastic cutting scheme was far from being in accord with the goals of the grassroots organizations—nor the goals of the Forests Forever initiative they *and* the mainstream groups had fought so hard to pass. It was, in fact, an egregious betrayal and marked, for many activists, a temporal boundary between third- and fourth-wave environmentalism.

Subsequently about 30 organizations, all of which had been shut out of negotiations, formed an ad hoc alliance called the California Forest and Watershed Council. The council asked for a hearing to present its position, but it was ignored by both sides in the negotiations (which by then had been joined by state legislators). At that moment 30 organizations, containing hundreds of committed activists and supported by thousands of two-figure philanthropists, unwittingly joined the fourth wave. "By excluding these interested parties from the process the Club damaged [its] credibility with local experts and created a climate of distrust between the Club and small active environmental groups," said

Sierra Club member David Orr, who attempted to persuade club leaders to broaden negotiations by allowing grassroots representation. "This is counterproductive for our movement. Big groups should be assisting, not resisting the small ones."[12]

Orr believes that the major beneficiaries of the accords were the politicians. "They carried the legislation, which they didn't have to write or amend. It was all done for them so that whatever the outcome they can plead innocent and stand for re-election without having angered either side."[13] The Sierra Club was been roundly criticized for its exclusionary politics and accommodations with big timber and has yet to take a public stand against clearcutting in national forests. The only heartening result of the Sierra Accords is a nascent reform movement within the club. It may yet produce an insurrection leading to the largest splinter in the club's history.

In the east too there were rumbles of dissent within the Sierra Club. In 1990 the 40,000-member Atlantic Chapter led by a feisty Brooklyn activist named Margaret Hayes Young supported several other organizations advocating an end to logging on public lands in the Rocky Mountain region. The national managers and the board informed Young and her associates that their actions were contrary to club policy. When Young persisted, national officials threatened to remove her and other local officers and to suspend the chapter. In the meantime they cut her budget and forbade her to contact Sierra Club donors in her region (New York and environs).

The Sierra Club has always prided itself in being a grassroots-driven, democratic organization. So when word spread through the club's e-mail network, members and member groups in Illinois, Indiana, and Montana expressed support for Young and her chapter. As club leadership continued compromising on wilderness and forest issues, a self-described "fundamentalist" insurrection began inside the club. In the summer of 1991 David Orr, who lives in Davis, California, formed the Association of Sierra Club Members for Environmental Ethics. ASCMEE, Orr declared, would be "dedicated to restoring the Club to its rightful place at the cornerstone of the environmental movement. Following the spirit of John Muir, ASCMEE will work to make the Club as environmentally ethical, aggressively pro-wilderness, and biocentrically visionary as possible. . . . We're not taking the club over," he promised his followers on many occasions, "we're taking it back." Orr suggested as the slogan of the new group, "Not blind opposition to compromise, but opposition to blind compromise." It was a paraphrase—or perhaps a parody—of one of the club's own mottoes: "Not blind opposition to progress, but opposition to blind progress."

In 1991 ASCMEE's picketing of the club's national board meeting incited the board to object to use of the words *Sierra Club*. It issued a cease-and-desist order against the group, whereupon the dissidents promptly changed the name to The John Muir Society.

The following year the Downstate Office of the Illinois Chapter of the Sierra Club drafted a full-page ad expressing dissatisfaction with the way the club was managed and ran it in several major newspapers. "Let us be clear," read the ad, "we are not ungrateful for the many honorable and vital accomplishments of the Sierra Club. But its endless willingness to make concessionary deals creates the very 'political reality' that the club cites as the primary source of its devotion to compromise. Belief in an antagonistic political reality, in effect, becomes a self-fulfilling prophecy."

By February 1993, sensing a major rebellion in the ranks, club leaders convened a meeting of forest groups in Washington. "As soon as we arrived they voted not to allow discussion of zero cut," recalls Jim Bensman from the Illinois Chapter of the Club. "Zero Cut" the slogan of the National Forest Council, a Sierra Club splinter, calls for absolutely no logging on public land. Instead "they adopted a new forest policy calling for an end to logging on *virgin forests* on public and private land." It wasn't enough for Bensman and fellow dissidents, who wanted *all* public forests protected and were incensed that the findings of conservation biologists were not being translated into club policy.

Believing that "the club's saving grace is its democratic structure," the rebels gathered signatures from members to place a referendum in support of a new forestry policy on the ballot of the club's next election. The referendum stated that "The Sierra Club supports eliminating logging for wood and fiber production of all public forests. This should be accomplished through 1) immediately ending commercial logging [of] all roadless areas, old growth, virgin and ancient forests, and other sensitive areas, and 2) phasing out, as soon as practical, logging of other public forests in conjunction with reduced consumption of wood and fiber, a transition to diverse and sustainable rural economies, and the promotion of sound forestry practices on private land." More than enough signatures were gathered. However, the referendum the national board placed on the ballot read "Shall existing Sierra Club forest policy be retained as is, and not changed by amendment as proposed by the petition?" Because of the confusing phrasing, a "yes" counted as a vote *against* the real petition for a revised club policy.

A few weeks before the election *The New York Times* reported that the club's chief northwest political director, William Arthur, had been logging a small timber track on his own land outside Cusick, Washington. Club dissidents used the exposé to illustrate how insensitive and corrupt some environmentalists could become, but the club's senior staff in San Francisco rose to Arthur's defense. He was not, after all, cutting virgin forest, nor was he clearcutting (only removing 70 percent of the timber). And, they added, it was his own property.

Although Arthur's actions aided the dissidents' cause, their referendum lost by 59 to 41 percent. (One dissident was, however, elected to the board in that election.) The vote was not a victory but it sent a strong message to a leadership all too willing to compromise with the timber industry that there was an incipient democratic uprising in the ranks. Because it happened at a time when club membership was declining disastrously, the board paid some attention. Still, the compromising didn't stop, and the national leadership went out of its way to keep ASCMEE from building a majority base in the club or openly opposing legislation or issues that headquarters had decided to support.

Whether the rise of ASCMEE will lead to the formation of new splinters, as insurrections in the club have in the past, remains to be seen. There is no sign that the hostilities have abated. In 1994 the club's top management threatened to cut off computer access to members using the e-mail network to insult club leadership. "We used to wake up knowing who the enemy was," lamented Executive Director Carl Pope in September. "All too often lately we seem to be treating each other like the enemy."[14]

Michael Fischer, former director of the Sierra Club, believes the club has no alternative but to oppose the rebels. "Were club directors to adopt the approach that 40 percent of the members voted for, the club lobbyists would lose the access it has in Washington. Committee staffers would simply refuse to meet with them because their own credibility would be affected if they did. Sierra Club national leaders know that they can't just walk into Congress and say no more clearcutting. So we are stuck with the incremental approach, which we hope will lead to slow progress in the halls of power. The problem is the incremental approach lacks the ability to stir people's souls, to get them angry and fulfilled."[15]

In the meantime, similar uprisings have occurred in the National Audubon Society. When the society invited Waste Management Corporation President Philip Rooney to join the NAS board in 1991, Florida Audubon members protested that "Audubon is for sale" for $60,000, the size of WMI's donation that year. The grant was to help finance an Audubon campaign for strengthened regulation of almost 200 tons of oil-field waste created by drilling companies every year. The waste was to be managed, of course, by Waste Management, Inc.[16]

The controversy over Waste Management's support of Audubon led to an attempt on the part of the Florida-based Save the Manatee Club to secede from the society. The director of the Manatee Club was summarily fired by Audubon, and locks were changed at the club's local office.

NIMBYs of Many Colors

You're beginning to see the emergence of grassroots citizen-based movements that I think are going to be the future of American politics in the '90s and into the next century.

Ralph Reed, The Christian Coalition

Environmentalists must be fierce and compassionate.

Terry Tempest Williams

Native communities are focal points for the excrement of industrial society.

Winona La Duke, Seventh Generation Fund

The American environmental movement has never been truly American. While the first three waves of environmentalism persuaded almost all Americans to admit that, deep down, they were environmentalists, the leadership, membership, staff, inspirational philosophers, and (until very recently) most of the activists have been well-bred, well-educated, white, mostly Anglo-Saxon, and mostly male. As social commentator Richard Rodriguez recently commented, "American environmentalism is a secular religion of the white middle class."[17] This fact too is about to change.

Anti-toxics populism and the struggle for environmental justice have brought people of color into high relief in the environmental movement. They network constantly and have strengthened the whole NIMBY process, creating an environmental vigilance that protects parts of the country in ways federal agencies never could. It has become virtually impossible to build a dam, open a landfill, fire up an incinerator, drill a deep injection well, or store nuclear waste without an immediate, well-organized response from surrounding communities, backed by solid understanding of all relevant laws and regulations. More important perhaps, such communities exhibit a willingness to fight not only the polluter but any national organization prepared to undermine the position. The environmental plight of Native Americans and their recent response to it offers a good case in point.

Since the arrival of Columbus about 2,000 North American native communities have been eliminated. About 700 remain—200 in Alaska, 80 in California, and the rest scattered around the continent in small reservations created by treaty or forced migration. As land available for grazing, logging, mining, and hazardous-waste disposal became increasingly less available in the lower 48 states, white Americans began to look to the marginal land their ancestors had set aside for native communities in the nineteenth century. At the time they had seemed like the least desirable lands on the continent, being mostly scrub forest and desert. Beneath many reservations, however, lay a wealth of resources. One-third of all the low-sulfur coal in the western United States is on or under Indian land. Two-thirds of all known uranium ore is under Indian land—a fitting irony, as all nuclear weapons detonated in the southwest have been on or under Indian land. Now Indian land is being eyed by the waste-management industry, which has lost hundreds of disposal sites around the country to the force of 10,000 NIMBYs. Of the current 18 "monitored retrievable nuclear storage sites" 15 are already on Indian reservations.

Native Americans, whose recent oppressions have been largely ignored by mainstream organizations, have had to become environmental activists in their own defense. The battles and lawsuits fought by native communities—at least one against a traditional environmental organization (Mountain Utes against the National Wildlife Federation)—are emblematic of fourth-wave activity. They are independent expressions of the need for protection of community health and direct action against polluters and are based on the conviction that neither government nor the traditional environmental movement can be relied upon to deliver environmental justice.

Lois Gibbs' strategy of attacking pollution neighborhood by neighborhood, the new forest movement's moves to protect trees grove by grove, and Native Americans' defense of their home environment, reservation by reservation, are all expressions of a new civil authority that will define the fourth wave of American environmentalism. That authority will be assumed by the grassroots as the federal government, with the occasional complicity of the mainstream movement, further abdicates its traditional and statutory authority over the environment. Yet the grassroots must face a monumental challenge, for no movement can proceed very far without a binding ideology. Creating a new ideology from the stew of ecological and social philosophies that surround contemporary environmentalism will be among the fourth wave's first and greatest tasks.

In Search of Manifestos

Me and nature are two.
Woody Allen

Now the chemistry of the atmosphere—the molecular structure of carbon dioxide—transforms Thoreau from an interesting crank to an urgent prophet.
Bill McKibben, The End of Nature

Contemporary environmentalists have yet to articulate a coherent social philosophy, and most American social philosophers have failed to integrate the science of ecology into their thinking. Environmental populists appear to have no ecological philosophy, relying instead on Marxism and other shopworn social ideologies for their argument against polluters. Thus American environmentalism remains fuzzy at the edges, ill-defined and semantically confusing in all its forms. It has always been and remains, in the words of former *New York Times* environmental reporter Philip Shabecoff, "a revolution waiting for a manifesto."

What is *the environment*? What exactly is *environmentalism*? How many environmentalisms are there? How many do there need to be? How many can exist before the term itself becomes meaningless? There seem to be as many answers to each of these questions as the number of people asked. And are the 87 percent of the American people who call themselves *environmentalists* really environmentalists, or are they just a lot of people who want potable water flowing out of their taps? Is Fred Krupp an environmentalist, or an environmental mediator? Is Dave Foreman really an environmentalist, or a beer-drinking redneck who happens to love grizzly bears?

Some conservationists are so frustrated by the multiplicity of terms and ideologies that they prefer to consider their movement totally separate from the environmental movement; Dave Foreman is among them. "There are really two movements," Foreman says, declaring himself to be part of "the *new* conservation movement"—itself a confusing designation, as he and most of its adherents are actually *preservationists*. Foreman differentiates his movement from the *old* conservation movement as well as from the present-day environmental movement, which, he asserts "is primarily concerned with human health. They are separate movements, and they should remain separate. Their agendas aren't conflicting, but they are different enough to require different names and different initiatives."[18]

As green consciousness deepens in some parts of the world, the potential for a truly ecological culture emerges. But a genuine shift requires a revised, if not completely new world view based primarily on ecological principles. The problem with ecological views and their attendant agendas is that no matter how rational and appealing they may be, they are eventually subsumed by more compelling ideologies, which are rarely ecological. Theologian Thomas Berry put it this way: "The energy evoked by the ecological vision has not been sufficient to offset the energy evoked by the industrial vision. . . . We need an ecological mystique rather than an industrial one." [19] Given the class and culture of the traditional American movement and the enormous diversity of the new movement, that will be no easy task. What is required is a radical change in values for every sector of our technological culture.

Challenging the inertia of obsolete world views is a difficult task. Protectors of political and religious ideologies—particularly capitalism, Christianity, and Marxism—have resisted efforts to integrate an ecological framework into their philosophies. But in recent years, as environmental consciousness has expanded, the adherents of even these faiths have been forced to relax their guard, and to reconsider their rejections of an ecological perspective.

There is no scarcity of new ecological ideologies. A dozen or more of them exist, each with its cadre of articulate advocates and ardent followers. But a dozen separate and conflicting environmental ideologies cannot form or inspire a movement. American environmentalism needs a synthesis of disparate and apparently antithetical ideas—an indigenously American ideology that can be comfortably embraced by major institutions, academic disciplines, economic philosophies, and religions. Very slowly, and in a piecemeal way, such a synthesis appears to be forming.

It is taking shape at an exciting, though at times confusing point for chroniclers of contemporary thought. "How can ecology, which aspires to be a unifying social vision grounded in an ecological outlook, express the concerns of workers, people of color, feminists and peace activists who are focused on exploitation, racism, sexism and militarism?" Howard Hawkins, a Vermont carpenter active in the U.S. Green party, asked around the time of Earth Day 1990. "How can ecology integrate these concerns with those of environmentalists focused on ecological destruction? Who will be the key social agents of ecological and social transformation? How should the movement organize itself? What should be its strategy of social change?"[20] Many of those questions were addressed during the four years that followed, particularly at events like the People of Color Summit in October 1991.

Hawkins' questions point to the need to integrate the problems and goals of diverse social movements into an overall philosophy of environmentalism capable of drawing adherents from widely disparate groups. Yet, if feminists, workers, and civil rights activists have seemingly divergent world views, so also do the various strands of the environmental movement itself. Biologists, romantics, social ecologists, bioregionalists, Native American ecologists, spiritual ecologists, greens, deep ecologists, and Marxists—all embrace their own ecologies. Even bioregionalists disagree among themselves, and green parties, small as they are, are often divided into two or more wings. Feminist ecologists and ecofeminists, themselves divided over minor points of philosophy, look on in horror as deep ecologist David Foreman and social ecologist Murray Bookchin rail at each other over what seem like surmountable differences.

Another unanswered, and crucial question (in spite of its seemingly abstract philosophical nature) is that of the relationship of the human individual, and human society, to the environment. Would an environmental ideology see humans as an integral part of the environment or would it regard the environment as something that surrounds humans without quite including them? Are we wrapped by the environment or somehow above it? Is the environment our gestalt? Or are we and our created environment the planet's gestalt? Is it to be "man [*sic*] *and* nature" or "man *in* nature"? One way we perish, the other way we survive, say various ideologies. But which is which? However you see it, it seems clear that we need an entirely new perspective. Name it *ecocentric, biocentric,* or whatever you like. The old paradigms, life styles, and ideologies of industrial culture must be challenged, and eventually replaced.

Almost daily the environmental imagination is bombarded with the new ideas, scientific findings, and paradigms that may one day constitute an authentic ecocentrism. Among the most powerful influences are bioregionalism, deep ecology, social ecology, feminist ecology, and dozens of early Native American ecological ideas and practices unearthed by cultural anthropologists. These new and rediscovered teachings challenge not only the anthropocentric canons of western philosophy but also the dominant, and fuzzy, paradigms of modern environmentalism, which, almost all fourth-wave thinkers agree, have failed to achieve their intended purpose.

Ironically, the formative ideas of American environmentalism—particularly those of John Muir, Aldo Leopold, and Rachel Carson—are not dissimilar from the positions expressed today on the radical fringes of the movement. Carson, Muir and Leopold, whose books gather dusk on the elegant cherry bookshelves of Beltway environmental organizations, would surely not recognize the occupants of these offices as their intellectual and spiritual descendants. They would be more likely to look to Bill Devall, Carolyn Merchant, Murray Bookchin, and John Seed—leading ideologues of the fourth wave—as keepers of the faith.

The new eco-ideologues, like their progenitors, call for preservation, not conservation of wilderness or resource management by government agencies working for extractive industries. But, unlike their ancestors, they see humanity as part of nature, not as master of it. They call upon science not to prove or defend their arguments or scenarios, but to demystify nature and explain ecology. Their ideas contain the beginning of a synthesis, but the sentiment isn't new. Rachel Carson, in a speech made as she was dying of cancer in 1964, laid out the ideological foundation for the next generation of environmentalists. "What is important is the relation of man to all life. This has never been so tragically overlooked as in our present age, when through our technology we are waging war against the natural world. It is a valid question whether any civilization can do this and retain the right to be called civilized. By acquiescing in needless destruction and suffering, our stature as human beings is diminished."[21]

Barry Commoner addresses the same relationship by neatly dividing the human world into two competing spheres, the ecosphere and the technosphere. "If the technosphere is ignored," he says, "the environmental crisis can be defined in purely ecological terms. Human beings are then seen as a peculiar species, unique among living things, that is doomed to destroy its own habitat. Thus simplified, the issue attracts simplistic solutions: reduce the number of people; limit their share of nature's resources; protect all other species from the human marauder by endowing them with 'rights.' This approach raises a profound, unavoidable moral question: Is the ecosphere to be protected from destruction for its own sake, or to enhance the welfare of the human beings who depend on it?"[22]

If American environmentalism is to remain a vital force in the next century it will be because its leaders return to the fundamentals of the original movement, recoup some of the passion lost during the previous three decades, and create a dynamic working vision of environmental recovery. For this reason, in its early stages the fourth wave will need to be as ideological as it is active. A driving philosophy for a rejuvenated movement will be forged from ideas contained in the many ideologies currently competing for the environmental imagination.

Fortunately Americans are an eclectic people. Out of the potpourri of ideas described in the next few sections will emerge a new environmental imagination. It is impossible to predict exactly what form it will take, but from the common ground around the competing philosophies one can safely pick up the following tenets and characteristics:

- Anthropocentrism will become passé.
- Humanity will be returned to nature and will no longer be considered the crowning achievement of evolution.
- All existing economic orthodoxies will be challenged.
- *Sustainability* will be clearly defined, well understood, and taught in school
- Clean air, clean water, and arable land will be considered basic human rights.
- Environmental justice will be litigated as a basic right and tried as such before federal and state supreme courts.
- Toxic pollution will change from a misdemeanor to a felony and be punished as such.
- A land ethic will be taught in nursery school.

Deep Ecology

If a war of races should occur between the wild beasts and Lord Man, I would be tempted to sympathize with the bears.
John Muir

Of all the new ideologies to surface during the past 25 years, few have challenged the environmental imagination as deeply as deep ecology. The term was first used in a 1973 article written by Norwegian philosopher Arne Naess as a counterpoint to what he saw as the narrow, utilitarian thinking and "shallow" ecology of the contemporary environmental movement. Naess defined shallow ecology as "the fight against pollution and resource depletion, the central objective of which is the health and affluence of people in developed countries."[23] A worthy struggle, but not enough, said Naess.

Deep ecology, he hoped, would lead environmentalists to "ecocentric identification" with forests, mountains, rivers, wildlife and deserts. Rather than speaking *for* nature deep ecologists see themselves as nature speaking for itself or, as Australian deep ecologist John Seed puts it, " 'I am protecting the rain forest' develops to 'I am part of the rain forest protecting itself. I am part of the rain forest recently emerged into thinking.'"[24] The American seed of this ideology was sown earlier in the century by Aldo Leopold, who implored his readers to "think like a mountain." From identification with nature flows *biocentrism*—as opposed to *anthropocentrism*. Humanity is no longer seen as the crown of evolution, the goal of creation, but as a part of it. Man-and-nature becomes humanity-in-nature.

To some degree the popularity of deep ecology was inevitable. It made an ideal counterpoint to the centralized, bureaucratic environmental activism that in recent years has actively restrained radical thought. By challenging reform environmentalism's unwillingness to criticize industrial society, technological culture, and capitalism, deep ecology attempts to set a new agenda and strategy for environmentalism. Deep ecologists are especially critical of third-wave environmentalism, which they see as narrowly rational and overly focused on economics and public policy. "No recognition is given to ecological sensibilities," complains American deep ecologist Bill Devall, "or to the necessity of providing an ecocentric critique of industrial society, or of exploring the 'wild self.' "[25]

It's all very inspiring, but it's also puritanical, insulting, and in its contemporary absolutist form, it is futile. "The rads foolishly try to lure the libs further out by chastising them for the inadequacy of their vision and commitment," complains lifelong eco-activist Stephanie Mills. "This puritanical approach, massively unappealing, guarantees only a tiny following of visionaries and masochists, folks who function best when guilty."[26]

Throughout its history the mainstream environmental movement *has* pursued a cautious and some might say narrow or "shallow" ecology. But to say so publicly and slight their world view as unevolved is politically counterproductive, and it's naive. For a mainstream organization departing from the environmentalism of its benefactors would be economic suicide. Why ask them to destroy themselves? Instead of railing against their fellow enviros fourth-wave eco-philosophers should proselytize a deeper ecological world view to philanthropists,

Another unfortunate result of the tactics used by deep ecologists has been the alienation of working men and women, particularly rural blue-collar workers. Not only do deep ecologists moralize about consumerism and believe in reducing human standards of living—neither of which are popular notions among people living close to the edge—but they tend to sanctify the species workers see as threatening their standard of living. The way to a logger's heart is not through the nest of a spotted owl. It's through the home of a fellow worker who has been laid off by a subsidiary of Maxxam or Georgia Pacific at the conclusion of a clearcut. Realizing this, a few deep ecology forest activists have allied themselves with labor unions and attempted to recruit loggers to the cause of sustainable-yield forestry. They have even denounced tree spiking, which rallied more supporters to the Wise Use movement than any single tactic of deep ecological activism. Maiming mill workers is just not effective organizing; linking jobs to environmental justice is. Which leads us to Deep Ecology's leading competitor.

Social Ecology

Philosophers have only interpreted the world in various ways; the point, however, is to change it.
Karl Marx

According to the teachings of social ecology, the domination of nature by humans began with the domination of humans by humans. This perspective was first articulated by Murray Bookchin in a 1964 essay entitled "Ecology and Revolutionary Thought."[27] Before humans can be harmonious with nature, Bookchin argues, they must become harmonious with each other. Quite predictably, he calls for a stateless, communal society in which civil authority protects the environment through decentralized democratic processes. The most egregious of human dominations, of course, are those that develop along lines of race, class, and gender.

Although it is clearly the oldest of the ecologically based ideologies, social ecology received little attention outside Bookchin's small circle of anarchist followers until the European Greens used it as a philosophical base for their platform. In the United States social ecology was virtually unheard of until deep ecology provided a foil for debate. Bookchin rose to the occasion in 1987 at a U.S. Greens convention at Amherst, Massachusetts. He excoriated biocentric thinking as "just another centrism" and deep ecologists as "misanthropic" for advocating population reduction (no argument there). A widely publicized donnybrook between Bookchin and David Foreman ensued. After a testy exchange of insults and arguments the adversaries met, found enough common ground to be respectful, reconsidered some of their positions, celebrated ecological values, and agreed to disagree about the rest.

Bioregionalism

Nature is in it for the long haul.
Jim Harrison

Bioregionalists love to ask city folks if they know where their water comes from. If they say "from the tap," they can be lovingly berated for not knowing the topography of their watershed and, ever so gently, persuaded that they need to know and understand their place in the surrounding biota that is hidden from view by tall buildings and asphalt freeways.

Bioregionalism, described by Stephanie Mills as "a biological politics so decentralized and wholistic it all but defeats explanation," proposes the restructuring of the human community along natural boundaries. Rather than dividing a nation into states or a state into counties with arbitrary gridlines, the political subdivision and the watershed would be coterminous; that is political subdivisions would be natural subdivisions. The political boundary of a bioregional territory would not run, as it so often does presently, along a meridian or even a river. Rather, it would be drawn along the headwaters line of every stream that feeds the river, the divide; each division would, if possible, contain but one single natural ecosystem. Once restructured along such biologically rational lines, bioregionalists believe, society would naturally evolve toward decentralization and political decisions would be made (democratically, of course) in accordance with the ecological health of the biotic region. "People are an integral part of life-places," writes Peter Berg, founder of

Planet Drum, a bioregional think tank in San Francisco. "What we do affects them and we are in turn affected by them. The lives of bioregions ultimately support our own lives, and the way we live is becoming crucial to their ability to continue to do so."[28]

Although bioregionalism makes considerable sense in theory it seems hopelessly romantic at present. American bioregionalists have conducted some remarkable restoration projects, most notably in the Mattole Valley of northwestern California, which had been ravaged by roadbuilding and clearcut lumbering. As a nationwide vision of the immediate future, however, bioregionalism grows increasingly impractical as more and more rural populations become dependent on resource extraction for their economic well-being. People who move to rural areas aspiring to live and govern themselves bioregionally have no authority to levy taxes. They soon find themselves dependent on the remote government of a county or state for support services like fire protection, libraries, public education, police, and emergency medical attention.

Only in New Zealand (itself a small enclosed bioregion) is bioregionalism being attempted in earnest. Government planners there have divided the country into 14 bioregions for the purpose of managing a national environmental plan. The borders cross over the county grid arbitrarily imposed by British colonialists more than century ago. Green planners in Auckland hope that nonenvironmental services like road repair and fire protection will be handled in bioregions in the near future and that eventually regional governments in New Zealand will correspond with areas defined by natural boundaries.

In the United States, bioregional councils modeled on a few formed in California during the late 1970s have been created in the Hudson River Valley, the Kansas prairie, the Ozarks, Appalachia, and Cape Cod. A national gathering of council leaders has been held annually since 1984.

Feminist Ecology

Given the male dominance and masculine bureaucratic structure of mainstream American environmentalism, the emergence of ecofeminism or feminist ecology—two different but similar ideologies focused on the “feminine” side of nature—was probably inevitable. Through separate spiritual and political factions, ecofeminism addresses the role of gender in environmental degradation, the ecological contribution of women in human society, and the impacts of technological culture on women’s bodies and human reproduction. Believing that masculine tendencies and male dominance have led to ecological crisis, ecofeminists attempt to restore women’s power and influence to the political process. Women, they believe, can play a singular role in creating ecological harmony. “For women to act as agents of environmental change, they must be freed from the narrow male assumptions about appropriate gender behavior,” writes Joni Seager, “and they must be free to act without the threat of male violence. . . . The struggle to form an environmentally just and sound future is inextricable from the struggle for gender justice and equality.”[29]

Like social ecologists, politically oriented ecofeminists believe gender domination led directly to domination over nature. Bring one into balance, they believe, and the other will follow. Like deep ecologists, the spiritual ecofeminists teach that only personal transformation will bring humans into harmony with one another and, eventually, with nature. But there are many voices of ecofeminism—spiritual, cultural, liberal, and socialist. “Ecofeminism not only recognizes the multiple voices of women, located differently by race, class, age and ethnic considerations,” says philosopher Karen Warren. “It centralizes those voices.”[30]

In 1993 women attending a gathering of forest activists in Oregon interrupted the program without warning (with no resistance from men) to discuss their concern that the convention, in fact the entire environmental movement both mainstream and grassroots, was dominated by men. The result was the drafting of the Ashland Principles, six guides to gender interaction that have been circulated throughout the environmental movement. “Women are not caretakers and men are not aggressors. . . . Women and men are natural allies. . . . We are *all* leaders in the environmental movement.” The principles constitute yet another step away from a male-dominated movement, another manifesto of fourth-wave environmentalism.

Spiritual Ecology

God is Green.
The Archbishop of Canterbury

As environmentalism has developed its own sciences, philosophy, and world view it has emerged in many countries as a remarkably alien ideology. This is perhaps especially true in America, where it has flourished. Environmentalism confronts so many assumptions of American culture—not just obvious things like anthrocentrism and Cartesian dualism, but also the Judaeo-Christian nucleus of western culture, every article of faith of which is challenged by ecocentrist thinking. American ecologists have frequently blamed Judaeo-Christian thought for the sorry state of the world's environment—a critique that has led to fundamentalist tirades against the "pagan" religiosity of environmentalism. While it is true that the book of Genesis encourages humanity to subdue nature and hold dominion over all other living things, most versions of the Old Testament also ask us to replenish the earth—though ever so parenthetically. With the notable exception of Buddhism, none of the world's major religions teach a primary reverence for nature. Thus almost all great civilizations have lacked ecological values and principles.

Earth Day 1990 was celebrated in churches and synagogues across America. In some, like St. John the Divine in New York City, it was treated as a religious holiday. While the Christian right sees environmentalism as pagan, celebration of nature is encouraged by church leaders who espouse more pantheistic views than their predecessors. Theologians are scanning the scriptures for ecological wisdom, compassion for wildlife, and reverence for nature. Presbyterians recently issued a "theological understanding" calling for the church to become involved in the "ecological-social crisis." The Methodist church has issued policy resolutions on environmental stewardship, and American Baptists (northern convention) recently declared that ecology and justice are inseparable.

Progressive priests, ministers, and rabbis preach that surrounding ecosystems are part of a parish and that plants and animals are also members of the congregation. Once a year at St. John the Divine, the

Episcopal cathedral in New York City, that teaching becomes reality. A special service is held for all living creatures, to which people bring giraffes, sea turtles, oxen, birds, horses, elephants, and whatever other beasts can borrowed for the day. All year long as well St. John's parishioners are encouraged to be environmentally active.

As environmentalism penetrates religion, humans of all religious persuasions begin to see themselves, in every regard, as part of nature. Concern for "a threat to our species" is replaced by a determination to fight "threats to all species." "I am working to protect God's creation" becomes "I am God's creation working to protect itself." Though many mainstream environmentalists still regard such views as flaky New Age thinking—a futile quest for social models in nature—they are in fact about as Old Age as any idea can be. Their roots are deep in religious traditions that predate those associated with western civilization.

Native Ecology

. . . Integrity is wholeness, the greatest beauty is
Organic wholeness, the wholeness of life and things, the divine beauty
of the universe. Love that, not man
Apart from that . . .

Robinson Jeffers, "The Answer?"

The sense of the Native American land ethic is so utterly self-evident that it seems unnecessary to repeat it: *Humans must co-exist with other forms of life on the land and not destroy it.* To America's earliest settlers, however, there has always been more to environmentalism than a simple ethic about the land. A collection of humans living individual lives in harmony with nature is not enough. Community is also required. And community to Native Americans, in fact to most indigenous cultures on the planet, almost always includes the biota.

In September 1993, at an international gathering of environmentalists held at the Esalen Institute in California, Jeanette Armstrong stood up to introduce herself. Armstrong, an Okanagan Indian from British Columbia, looked at her colleagues one by one as she spoke. Most of them had already introduced themselves, when she said:

Listening to those introductions confirmed what I have heard and feared about whites. Every introduction was about yourself, not about your community. You seem to have no community connections.

If that's true, I'm sorry for you.

In our culture, we identify ourselves in relation to our group. We don't know who we are except in relation to our family, our community and the land that is our life. We don't know ourselves unless we know how nature works. Our language reflects the language of the birds and flowers and trees. I identify with my people and with the plants; and I am the river too. Our first law is the law of the natural world that gave us life. We cannot do anything that injures the natural world. We would be injuring ourselves.

You can't imagine the pain your society creates by breaking up our connections to our land and our community. Grand Coulee Dam flooded the land of our ancestors, and we were "relocated." Our children were taken away to be educated at residential schools, and they returned without any sense of community. They didn't know how to relate to each other, or to plants and animals in a loving way.

Your disconnections with *your* community ruins ours. I'm happy to be here with you, but mostly I feel grief for you. And I'm afraid for the world.[31]

The environmentalists who had joined Jeanette Armstrong that day sat silently for 10 minutes after she spoke. Perhaps when they left Esalen they carried with them to their own nations a new understanding of community.

Fourth-wave environmentalism has begun to see the melding of natural and industrial environments, rural and urban, wildlife and human environments into a singular concept. Community will mean all life within its border, the entire biota, be the border natural or arbitrary. Eventually all environments will be seen as one. There is only one biosphere on the planet. There is only one body of water, one habitat, and only one very thin mass of air reaching into the canyons, forests, cathedrals, factories, and lungs of every breathing creature. Through that consciousness fourth-wave thinking will reaffirm, and to the extent possible, restore the land-use ethic of America's earliest known settlers.

The Sustainable-Development Perplex

Growth is the engine of change and the friend of the environment.
President George Bush

Sustainability is not something to be defined but something to be declared.
Bert De Vries

Central to the ideology of fourth-wave environmentalism is the notion of *sustainability,* a term that defies definition. Ever since it was first used by Gro Harlem Brundtland in the now famous 1987 Brundtland Commission Report, it has been one of the most hotly debated concepts in contemporary American environmentalism. *Sustainable development* has received so much and such varied usage and led to so many gatherings in academia and government, that architect Sim Vander Ryn, who has attended many of them, dubbed it "the mantra that launched a thousand conferences."[32]

Despite the thousand conferences, sustainability and sustainable development remain two of the most overdefined concepts in contemporary western thought. The struggle to define them correctly has created a nexus of economists and ecologists; together they have formed a small industry that attempts to reconcile the seemingly irreconcilable goals and tenets of their disciplines. A definition can range categorically from the ecological to the physical to the economic, depending on the person creating it or the conference he or she is addressing.

Heartening to some, threatening to others, the concept of sustainable development seems bound to remain at the center of an ideological firestorm for the foreseeable future. Although most observers would accept Jon Roush's general observation that "sustainability describes a system in which human societies and natural systems [are] in an equilibrium that can sustain itself," when it gets down to details, one person's sustainability is another's exploitation.[33] Extractive corporations say they are practicing it; ecologists say they aren't even close. To a corporate economist sustainable development means development that will allow his company to remain in business forever. To an environmentalist it is development that will allow the earth to stay in business forever. To a deep ecologist it is virtually no development at all as we know it.

Symptoms of global malaise generate almost as much disagreement among earth scientists as they do among politicians. Global warming and ozone depletion are particularly troublesome for scientists because evaluations of their impacts are conjectural and based largely on theoretical projections. Even so, the following threats to sustainability are widely acknowledged to be undeniable.

• Human beings—only one of between 5-30 million species on earth—currently consume 40 percent by mass of the plant material produced each year by photosynthesis—the GDP, as it were, of the earth. The rate of increase of human use is about 2 percent per year. At that rate human consumption will double in 36 years.

• Thirty-five percent of the land on the planet is degraded, much of it irreversibly. Soil loss currently exceeds rates of soil reformation 10 to 1. Fifty-five percent of the tropical rainforests on earth have been destroyed.

• Species extinction rates currently exceed 5,000 per year. Compared to the previous known spasms of mass extinction, this rate is disturbingly high—10,000 times higher than pre-human extinction rates.

• Despite a quadrupling of economic output since 1950, the percentage of humans living in poverty has continued to grow, even in the industrialized countries of the North, where over 100 million are classified as poor.[34]

Developed countries and underdeveloped countries have completely different, often conflicting definitions of what constitutes sustainable development. The final definition will be hammered out in the councils of world governance and will affect the future of the biosphere as much as any other force devised by humanity. It is essential, therefore, to grapple with the semantics of sustainable development before we can develop plans or systems for ecological sustainability. Even the word *development* needs to be challenged. "What does it mean for a country to become more 'developed'?" asks Stephen Viederman, president of the Jesse Smith Noyes Foundation, a major environmental funder that has been grappling with the question for some time. What will third-world countries look like when they become more developed in a world that values sustainable development?

It seems certain that one arbiter of sustainability will be those environmental organizations with enough global reach and resources to watch economic activity worldwide and apply substantial pressure where it is needed. For this reason, the economic and ecological precepts accepted by mainstream American environmental organizations could have a considerable influence on the future of worldwide development and, thereby, the health of the biosphere. Given the drift toward third-wave tendencies of some of the most prosperous mainstream organizations, that is a disconcerting possibility.

Even more disconcerting is the President's Council on Sustainable Development (PCSD), formed June 29, 1993, by executive order of President Clinton. It is another administrative creation designed to assure the public that the U.S. government is committed to a sustainable future. Dow Chemical, which has committed itself to "becoming the premiere company in the practice of sustainable development," is represented on the council by co-chair David Buzzelli, considered by some to be a paragon of corporate environmental enlightenment. Dow joins Ciba Geigy, Pacific Gas and Electric, Chevron, S.C. Johnson, Georgia Pacific, Exxon, and Browning-Ferris on the PCSD. Ciba Geigy and Georgia Pacific are also listed on the Council on Economic Priorities' "Toxic Ten"—the worst environmental offenders in all industries except petroleum (The oil companies are excluded because if they were included all on the 10-worst list would be oil companies.)

Joining the corporate executives on the council are cabinet secretaries Bruce Babbitt, Carol Browner, Hazel O'Leary, and Michael Espy, who sit interspersed with environmental leaders John Adams (NRDC), Fred Krupp (EDF), Michelle Perrault (Sierra Club), and Jay Hair (National Wildlife Federation). People of Color are represented by the Reverend Ben Chavis of the NAACP and by Native American Ted Strong of the Columbia River Intertribal Fish Commission. Notably unrepresented are representatives from anti-toxic grassroots organizations like CCHW. "It isn't that we chose to be unrepresented," comments Lois Gibbs, executive director of CCHW. "We weren't invited. Yet we are on the front line of the jobs-versus-environment issue. And our folks are really looking at environment and economics at the hardest levels—the coal mines, the incinerators, and the polluting factories."[35]

The only hopeful sign from the first few meeting of the PCSD is that third-wave environmentalism is being rejected by at least half the members. The council, which has created several task forces to examine problems important to economics and the environment, split 10–10 over whether or not to add a task force on market incentives. The success of the council will hinge on whether environmentalists can move beyond the clichéd indictment that economic activity threatens the environment and convince the business representatives that environmental degradation threatens economic activity.

So far the most positive result of the sustainable development debate is widespread acceptance of the fact that the environmental crisis is not a single issue and that environmentalism is not a one-dimensional ideology. It is a complex admixture of many disciplines and ideologies, most of which have something worthwhile to say about sustainability. Continued disagreement over the meaning of sustainability will be central to the dialogue of fourth-wave environmentalism, which will hopefully find and implement a definition that protects economic well-being, nature, and the biosphere.

The Players

The environmentalists are going to have to look like the mob in the square in Romania before they prevail.

Frank Mankiewicz, Hill and Knowlton

The fourth-wave of American environmentalism will include most of the players currently in the game—mainstreamers, NIMBYs, lawyers and rads. But there will be a whole new roster of people who during previous waves of environmentalism either stayed out of the game for their own reasons or were kept out by the dominant players.

The Left

For too long the left has attacked, disregarded or given only lip service to the environmental movement.

Jeffrey Escoffier, Former Editor, Socialist Review

From its beginning the participation of the American left (new and old) in the environmental movement has been limited and cautionary. Many leftists, particularly new leftists of the early 1970s, were outrightly hostile to environmentalism, deriding enviros as apolitical bourgeois romantics and anti-urban, elitist, utopian Luddites who believed, as Pogo did, that "the enemy is us." Such thinking, leftists believed, shifted responsibility for pollution away from the real polluter, corporate capitalism, just as certainly as the population-control zealots at the Rockefeller Foundation shifted the industrial North's attention away from the real problem in the third world—economic and cultural imperialism.

On Earth Day 1970 leftists across the country excoriated greens for distracting the public from other pressing issues: the war in Vietnam, civil rights, third world sovereignty, prison rights, women's liberation, etc. By Earth Day 1990, however, a short green passage had been discovered in the writing of Friedrich Engels, helping some leftists find aspects of the environmental imagination they could identify with. "At every step we are reminded," said Karl Marx's foremost comrade and patron, "that we by no means rule over nature like a conqueror over foreign people, like someone standing outside nature—but that we, with flesh, blood and brain, belong to nature, and exist in its midst, and that our mastering of it consists in the fact that we have the advantage over other things of being able to know and correctly apply its laws."

By the 1970s, some elements of what is now identified as the New Left, followers of Murray Bookchin, and back-to-the-land communard "dropouts" had developed an ecologically oriented critique of American daily life. The urban wing formed communes in New York and other major cities, where environmental action took the form of garbage strikes; and some countercultural groups, like the Diggers in San Francisco, evolved into bioregionalists. The rural drop-out pretty much disappeared as a political force until the creation of Earth First! in the early 1980s, which sparked a new preservationist movement in the United States.

The rest of the American left eventually came to see the inseparability of peace advocacy, civil rights, and environmentalism, and even to question the western industrial model as a suitable substrate for socialism. "Can or should the Western level of consumption be the worldwide norm?" asked the neo-Marxist *Guardian* (Manchester, England) in an Earth Day 1990 editorial acknowledging that large-scale centralized planning had created "the environmental crisis in the East [Eastern Europe]." The same editorial cautioned that "green capitalist" was still "a contradiction in terms," but that it was "crucial to listen to voices speaking from the rainforests, villages, and encroaching deserts."

Some American leftists even began to call environmentalism a left sensibility and part of their agenda. Traditional environmentalists, however, were not so sure they wanted to be identified with the left. In a society that associates Marx with Satan and liberal Democrats with bolsheviks, it was safer for conservation activists and ecologists to remain as nonideological as possible, defining politics as electoral activity and political activism as recruiting through direct mail and lobbying elected officials. Above all, direct confrontation was to be avoided and—until the Cold War was over and socialism declared dead—capitalism was never to be derided or challenged.

Although a leftist critique of environmentalism continues, American Marxists and other socialists have found common ground, not with environmentalists so much as with environmentalism and ecology. *Ecosocialism* really exists. The Wise Use rage-rats are right: there are a few watermelons in the environmental movement—green on the outside, pink on the inside. That's what makes the movement American.

There is one disturbing aspect of the New Left (or should it be New New Left?). Described by Nathan Newman and Andres Schneiderman of the Center for Community Economic Research in Berkeley as "market leftism," it has, unfortunately, become an integral aspect of the environmental movement. Market leftism is the selling of social issues and problems as commodities by canvassers, telemarketers, and direct mail services to "the usual suspects"—lists of people who have written one or more checks to organizations addressing similar social problems. (See Chapter 2 for more on the use of direct mail.) "Market leftism has the same corrupting effect as market capitalism," say Newman and Schneiderman. "Those with more money have more power, and the poor and youth who have less money end up with little voice."[36]

As products of the market left, national environmental organizations have become little more than "fundraising machines and mailing lists that no one bothers to mobilize," according to Newman and Schneiderman. Only in the environmental justice movement has "the melding of environmentalism and community activism" created enough energy to sustain a lasting social movement. Such a melding will be part of the fourth wave.

The Greens

Neither left nor right but in front.
Slogan of the German Green party

In 1983 Die Grünen, the Green political party of West Germany, surprised the democratic world and inspired American environmentalists who felt shut out of electoral politics by winning enough votes in a national election (more than 5 percent) to gain seats in the Bundestag. Although Green parties existed in almost every country of Europe and North America it was the first time more than one politician running as an environmentalist had been elected to a national legislature.

Members and voters of European Green parties are drawn mostly from social democracies, socialists, communists, and other parties of the left. In recent years, however, traditional Greens have been vexed by the number of rightists, even fascists, who have declared themselves "green." Enough of them have joined some Green parties for there to be a rightist green vote of consequence, particularly in France and Germany, where a right-left division simply exacerbates the older conflict between the *realos* (realists) and the *fundis* (fundamentalists).

Greens have had mixed success in Germany, sliding into obscurity for an election or two, then returning again to prominence. In 1994 the party, which had moderated its tone and became more pragmatic in practice, emerged from the October elections as the country's third most powerful party. This prompted pundits to speculate that if the left-center Social Democrats won a solid minority block they could join Die Grünen to form a red-green governing coalition.

Germany was a predictable country for a Green ascendancy. Their hasn't been a wild frontier there to pillage for thousands of years. Most everyone is committed to stewardship, and environmentalism has virtually no political antagonists. Because of proportional representation, a party that gets enough of the votes cast anywhere in the country can earn representation in government. There are Green parties in every country of western Europe that has proportional representation. In Poland, the Czech Republic, and other former East bloc countries, where environments were ravaged by state-owned and operated industries, Green parties are also appearing.

The German Greens' strong showing gained other European Green parties the media attention they needed to grow and win elections. European Greens, anxious to escape the single-issue label, embrace a critique of technological culture, are outspokenly anti-nuclear, and oppose the policies and institutions generated by the Cold War. While Green parties and candidates have adopted wider political identities, a Green anywhere is still at heart an environmentalist, and environmentalism remains the driving force behind party platforms. Yet only through a multi-issue framework, Greens believe, can ecological balance be obtained through political processes.

Lacking proportional representation, it is unlikely that American Green candidates will soon be elected to national or even state offices, although a few have won county and town council seats. In 1980 Barry Commoner, inspired by Green victories in Europe, accepted the presidential nomination of the Citizens party. Unlike die Grünen, the Citizens party (perhaps because of its unfortunate abbreviation, CP) never grew beyond a coalition of small activist organizations and marginal parties. In 1984 Commoner abandoned the party and became environmental advisor to the Jesse Jackson Democratic primary campaign. Since then the American Greens have organized into over 300 Committees of Correspondence and begun running candidates—mostly in state and local elections.

Like all third and fourth political parties in the United States, the Greens have had great difficulty getting organized. Struggles over style, ideology, platforms, and strategy persist. Although the party has not divided into formal right and left wings as Green parties have in Europe, an anti-capitalist Left Green Network is a loosely organized tendency within the party. There are also Progressive Greens (liberal), and a New Age faction. Such a small party divided into three tendencies is unlikely to be a significant factor in either American electoral politics or fourth-wave environmentalism. Nonetheless, the candidates and campaigns of American Greens will help keep environmentalism alive as an issue in electoral politics.

True democrats

The trouble with Brandborg is that all he wants to do is run around the country trying to make democracy work.

A critic of Stewart Brandborg's leadership at the Wilderness Society—which fired him.

Green party politics may not noticeably affect American politics. Democracy, on the other hand, will be a major factor in the fourth wave of American environmentalism. Already reform environmentalists have successfully employed democratic processes and institutions to achieve a degree of environmental protection. In the process, however, they have created a far-from-democratic regulatory apparatus. Its overseers are bureaucrats—many of them aspiring to "higher office" in the corporations they are mandated to regulate. Even the limited congressional oversight written into most environmental legislation is inadequate for controlling career regulators, administrative law judges, and other apparachiks empowered to enforce environmental laws. Although provisions for public participation are written into some bills, all too often regulatory legislation bypasses or obstructs community control and input.

The failure of the Clean Air Act and its many amendments to achieve the air quality levels described in its earliest preambles provides an example of how weak and undemocratic reform environmentalism can be. The federal government's inability to protect air quality has stimulated states and metropolitan regions to take on the task. If Los Angeles ever has clean air it will be because local enviros stopped flying to Washington to lobby EPA and Congress and stayed at home to work with area residents, the Los Angeles Clean Air Coalition, the Labor/Community Strategy Center, and others to challenge the Los Angeles Air Quality Monitoring Board and the polluting corporations.

Authentic civil authority over the environment has been abdicated by government and compromised by the mainstream movement. It is therefore incumbent on the next generation of environmentalists to forge a new civil authority with a mandate to protect land, air, water, and biota. The task will require hard work and imagination. Most of all it will require a commitment to democratic processes.

Human Rights Advocates

As environmental advocacy becomes a global phenomenon and environmental activism expands in the developing world, human rights activists concerned with the right to organize, demonstrate, and speak freely have opened a dialogue with environmentalists. The murder of Brazilian rainforest activist Chico Mendes in the Amazon basin and attacks on other third-world environmental activists seem to have excited their interest. This is a hopeful sign, particularly as no existing international human rights charter mentions a healthy environment as a basic human right. Protecting environmentalists' right to speak out and protest around the world is a worthy and important objective. It should go without saying that environmentalists have the same rights as other legitimate political protesters. The real service that the United Nations Human Rights Commission, Human Rights Watch, and other such organizations can provide to the environment is to expand their own concept of basic human rights to include clean air, clean water, arable soil and a more encompassing definition of who is to be protected and from what.

"In Western civilization the concept of rights is basic to improving the workings of democracy," asserts John O'Connor, founder of the National Toxics Coalition. "The right to life is a precondition for all other rights." This most basic of all rights must, according to O'Connor, be incorporated into the environmental endeavor. "In coupling the right to life with liberty and the pursuit of happiness, America's founders clearly demonstrated their belief that there was more to living than mere subsistence. They also provided an effective means of combating the social and environmental devastation wrought by polluting industries. The right to life cannot be guaranteed if corporations can poison the soil, air, and water with cancer-causing chemicals. The pursuit of happiness is an empty promise for the six million children, predominantly poor, who may suffer learning disabilities because of permitted environmental lead levels. If life is a right then we must protect those elements that sustain it. Since drinking and breathing are essential to life, clean water and air are rights as well. What good are any rights—free speech, religion, or assembly—if we are not healthy enough or alive to enjoy them?"[37]

The ultimate advantage of transferring environmental issues to a human rights context is that it makes compromising almost impossible. It is much easier to negotiate the cleanliness of water in the abstract than it is when water becomes drinking water and the basic human right to healthy drinking water becomes the issue. Human rights advocates do not generally negotiate half measures. ("You may torture 5 percent of your population, a 50 percent reduction from the 10 percent you tortured last year.") Nor would a human rights oriented environmentalist allow gradual reductions of toxiphene in a community's well, in return, say, for the complete removal of dioxins. And just as a human rights charter would never permit one nation to sell its "right to torture" to another, a human rights environmentalist would not permit one polluting nation to sell some alleged or implied right to pollute to another (an idea that is being seriously discussed by American third-wave advocates).

An environmental/human rights nexus is imminent. The human rights community made the first move 1992 in the form of a proposed joint project between Human Rights Watch and NRDC. HRW might have picked a better ally, but it's a start. And discussions have been under way for about five years between the Sierra Club Legal Defense Fund and the U.N. Human Rights Commission over the drafting of a resolution declaring a healthy environment to be a basic human right. A coalition of human rights advocates and environmentalists could become the inspiring model for alliances between American environmentalists and other social activists—from whom the former could learn some valuable lessons.

Ecological Economists

Economics is an advanced form of brain disease.
Hazel Henderson, Economist

Until very recently economics and ecology have operated as separate disciplines. They are now being combined as the new field of ecological economics and will, in due course, draw academicians into the fourth wave of American environmentalism. Although viewed with some condescension by "real" economists and "resource" economists, the new field represents a sincere attempt by both economists and ecologists to acknowledge and contend with the effect they have on each other's discipline. Its chief tenet states that all economic decisions require a long-term environmental impact assessment and that all ecological analyses involving a human habitat or workplace must take economic effects into consideration.

Already ecological economists are grappling with some of the most troubling questions about sustainability. Is the surplus produced by a rapidly growing economy the only possible source of funds for environmental protection and remediation? Does a dynamic industrial economy inevitably create an unending need for environmental protection by consuming more energy, generating more exhaust, more mine tailings, and more waste? Like other modern economists, ecological economists will no doubt reduce such questions to equations, develop microeconometric models, and apply calculus to environmental issues.

A serious problem for ecological economics is that no one can decide whether as a discipline it should be subsidiary to economics or to ecology. The answer, of course, is neither. But that's not the way science is organized, particularly in academia. When ecological economics is offered at universities—or better still as a required course at all business schools or a mandatory course for graduate work in either economics or ecology—the curriculum and faculty will probably be managed by the economics department. In most western institutes of higher learning, unfortunately, economists still regard regression analysis as the pathway to truth and the free market as sacrosanct. Outside academia as well the problem will persist, for most free market societies grant large corporations the central role in managing economic life.

Americans, once they gave up the democratic control over concentrations of power, formerly exerted through regulation, also lost the ability to control the ecological decisions of corporations. In an era of counter-regulation it will be difficult, if not impossible, to regain enough control to force corporate management to adopt even moderate ecological measures. There must, therefore, be power to apply pressure on the economic establishment beyond the polite strictures of third-wave environmentalism. It isn't enough to advocate ecology; it must be asserted. The fourth wave of environmentalism will assert ecology as a code of civil authority, not only in economics, but in all disciplines—particularly politics.

Madeover Mainstreamers

Most of today's decisionmakers will be dead before the planet feels the heavier effects of acid precipitation, global warming, ozone depletion, or widespread desertification and species loss.

Our Common Future, *Report of the World Commission on Environment and Development*

Desperate attempts to restructure, re-inspire, and reposition mainstream environmental organizations are being made. Unfortunately much of the effort is aimed at image while little is being done to improve the vision, effectiveness, or response time of the nationals. In 1991, for example, the National Audubon Society, on the advice of San Francisco packaging designer Walter Landor, attempted a complete makeover. In random focus interviews Landor learned that people preferred Greenpeace to Audubon by a wide margin, a finding that played directly into the hands of President Peter A. A. Berle (who had come into office in 1984). Landor's advice, which Berle received enthusiastically, was to transform Audubon into a forceful, politically active advocacy group—not an Earth First! or even a Greenpeace, but sort of a Greenpeace Lite.

"We won't be parachuting off bridges," said a staffer in the New York headquarters in introducing Audubon's new look. "But we want to become more like Greenpeace. We want to be much more than birders." Berle announced that there would be "unlimited opportunities to plug your energies into the Audubon machine." On his orders, the word "oldest" was to be dropped from all promotional materials and replaced by "one of the nation's largest and most effective. . . . " Optimistically, society officials predicted a membership jump to two million in five years, a very ambitious goal given the 500,000 who were signed up at the time. It was a bold, but in the end pathetic attempt to turn late middle-aged birders into activists. In the process flocks of birders quit the society and a small group of board members engineered a coup d'état intended to dispatch Berle back to Wall Street by August 1995.

National Audubon is not the only national reconsidering its future. With a generous grant from the Ford Foundation, NRDC spent much of 1994 reevaluating its mission. The Sierra Club, plagued by chronic budgetary shortfalls, did a lot of soul searching during the same period and cut out a few programs. In spite of similar efforts by most nationals, there is as yet no sign that any mainstream organization will close down or radically trim its Washington office and refocus its attention on winning back some of the grassroots support lost to the Wise Use movement. Until they do so they will continue to court irrelevance.

However attenuated their power and prestige, most of the organizations forming the mainstream of the environmental movement will survive in some form, although they will no longer *be* the mainstream. Most will continue to perform the functions they were created to perform, and they will do so competently. They are professionals. The question is whether those functions will be relevant to an adequate defense of the biosphere in the postmodern era.

Philanthropy Reconsidered

The direction taken by American philanthropy in the years to come will be critical to the future of environmentalism. Will the movement continue to receive about 2.5 percent of total American giving? Or will environmental donors, large and small, find more pressing needs for their charity and desert the cause? Contributors are a fairly fickle lot. If present trends continue, however, overall giving for environmental causes will rise gradually relative to total philanthropy, with the 25 largest organizations continuing to receive about 70 percent of the total. The 10,000 smallest organizations will compete for the scraps, as they have for the past 25 years. If the gross contributed remains constant only a major shift in its distribution will enable grassroots and fourth-wave organizations to receive adequate funding.

Even though most mainstream organizations are presently suffering revenue losses, their combined portion of the total pie has not changed significantly over the past decade. During their period of massive expansion in the 1980s, as they rushed to become all-purpose organizations and compete with one another for money and memberships, the nationals became clones of each other. They added projects constructed around "designer issues" that quickly became redundant. If money was available for an issue, a region, a crisis, or a species (preferably a furry mammal), every fundraiser worth his or her salary put out a proposal. "If a foundation had a large interest in Alaska and a lot of money," quipped Former Wilderness Society Director Bill Turnage, "you definitely had a large interest in Alaska." A new project was funded, a new department was formed, a few specialists were hired (or raided from other groups), and organization A began doing what organization B and C were already doing.

When the officials of W. Alton Jones Foundation declared in the early 1990s, for example, that they saw grassroots organizing to be the wave of the future, mainstream enviros became grassroots organizers—not, it turned out, very successfully. When environmental justice became hot, everyone had to have an environmental justice program. The result was six similar programs, all doing the same thing, and none of them doing it very well. The aggregate effect is what Pew Charitable Trust's Josh Reichert calls "growth at the sacrifice of focus. . . . a tendency to expand the thematic agenda to a point where one organization is dealing with 30 or 40 problems at the same time. Staffs are highly diversified and spread out over a wide terrain." From a management viewpoint, the organizations began to look like runaway trains.

Faced with shrinking philanthropy, Reichert predicts, organizations will have to change. If the entire mainstream movement has to downsize, as he believes it will, leadership will have little choice but to reevaluate what their organizations do best and do it, eliminating redundant or ineffective programs. In an ideal world the reorganizations would be coordinated among the mainstream organizations. The Green Group might even meet to divide up the mainstream pie, much the way crime syndicates divide up territories and enterprises: "We get the numbers in Brooklyn. . . . You get the horses in the Bronx" could become "Sierra Club takes book publishing and eco-tourism, the Wilderness Society handles Washington lobbying and protection of public land, SCLDF litigates in state courts, NRDC takes the federals." Such a meeting—alas unlikely to occur—would help eliminate the wasteful redundancy of effort in organizations that have been competing with one another for two decades, often viciously, for money, size, and prestige.

"It's interesting to note," says Reichert, "that the organizations that are growing and being the most effective in their work are the ones with singular objectives like the Nature Conservancy. They do one thing and they do it well." Reichert acknowledges that it takes vision and courage for the executive director of a large organization to go to his or her board and say "we have to cut programs." But sometimes it simply has to be done.[38]

The Sierra Club may lead the way. In October 1994, after a tense emergency board meeting, the directors announced a narrowing of the club's focus to two goals—protecting wild lands and preventing pollution. Instead of 63 national committees there would be 6, and many staff members were laid off. Others nationals may follow suit.

The problem of the disproportionate (70 percent to 30 percent) funding ratio between mainstream and grassroots movements remains however. A 20-point shift of, say, about $200 million out of the mainstream and into the grassroots movement would change the complexion of the entire environmental community. The mainstream would be forced to cut back even further, and the grassroots would blossom. Money would be more frugally spent, philanthropists would gain more bang for their buck, and environmentalism just might come alive again.

That process could be started by foundations. Some small- and medium-sized foundations have already made the shift, denying funds to regular grantees like EDF and NRDC and giving the money instead to such groups as the Citizens' Clearing House for Hazardous Waste, the Oregon Natural Resources Council, and Ozone Action. Not until large endowments like Pew, Rockefeller, Alton Jones, and Ford significantly alter their giving, however, will the shift have a noticeable effect.

In the end, of course, the future of the movement depends on the public's perception of the need for it. If industry can prevent major spills and disasters while their public relations mavens continue their greenwashing campaigns, it could become increasingly difficult to raise money for environmental advocacy. "Without thick particles floating in the air, without rivers bursting into flames, the issue is more difficult to perceive," observes Philip Shabecoff. "This is particularly true with an issue like global warming, which you not only cannot see but which will not affect the Earth for decades. How then do you get people to make sacrifices?"[39]

A Movement Reinvented

The hard political path is the only workable route to the soft environmental path.
Barry Commoner, Earth Island Institute

The ghosts of Muir and Pinchot still wrestle for control.
Stephen Fox, John Muir and His Legacy

I once had the pleasure of introducing Dr. Robert Bullard to a book convention. As a sociologist at the University of California at Riverside, and more recently at Clark University in Atlanta, Dr. Bullard has spent most of his professional career researching the travesties of environmental injustice . His findings, documented in three books, have fueled the burgeoning nationwide movement for environmental justice.[40] Thus no one was surprised when he stood before his audience, dressed in traditional African garb, and spoke eloquently about American communities of color that had been carefully singled out by the hazardous-waste management industry as locations for landfills, toxic waste dumps, deep-well injection sites, and incinerators. The surprising part came at the end of his speech, when Bullard spoke passionately about endangered species of plants and wildlife, about the rainforest, the need for wilderness, and the freedom of whales. It was a heartening sign of reconciliation, at least as significant as the speeches of white environmentalists who ended their addresses with impassioned calls for environmental justice.

I was reminded of Robert Marshall, the New Deal era conservationist and social activist who co-founded the Wilderness Society. Throughout his tragically short life Marshall lobbied for the set-aside and protection of public land, which he called "democratic wilderness." It should, he believed, be accessible to *all* the people who owned it—European Americans, African Americans, and Jewish Americans. (The latter two groups were barred from national parks at the turn of the century.)

Marshall, the son of a prominent Washington lawyer, graduated from the Syracuse University School of Forestry (which his father had endowed) and became an inveterate wilderness backpacker and defender of the early U.S. Forest Service. He believed that social liberation and wilderness preservation were inseparable goals and that creating public reserves for the elite would simply exacerbate the injustices of American society. The back country, he said, should be completely accessible to "the ordinary guy." It was not a popular concept among traditional conservationists, who feared that if "the multitudes" were brought into the wilderness, the wilderness would be "reduced to the commonplace."[41]

Marshall died in his sleep November 11, 1939, on a train from Washington to New York, some say of fatigue after too many long hikes and long battles against intransigent bureaucracies. His $1.5-million estate became the endowment that supported the Wilderness Society for years to come, and the vast Bob Marshall Wilderness area in Montana was named after him. But his most significant legacy is the long-dormant seed he sowed for a broad synthesis of social and environmental agendas. As social historian Robert Gottlieb notes, "Over time Robert Marshall's life and ideas began to undergo reinterpretation, with the suggestion that his love for wilderness had really been a separate concern. With his death, the 'people's forester,' whose life's mission had sought to link social justice and wilderness protection, would become an ambiguous historical figure representing environmentalism's divide between movements, constituencies and ideas"[42]

One cannot help feeling, that if Bob Marshall had lived a longer life things might have turned out very differently for the American environmental movement. Few environmentalists of his era understood the nexus between their movement and others in American society, and even fewer seemed willing to learn from the successes and failures of other movements.

The civil rights movement, for example, presents some obvious object lessons. It was not national organizations or their Washington offices that ultimately brought civil rights to America's racial minorities. They were won by direct confrontation, sit-ins, and "illegal" demonstrations led by leaders like Medgar Evers and Martin Luther King, Jr., who grew impatient with the intransigence of the federal government and saw clearly that civil rights and other contemporary movements were inextricably linked. King deliberately lifted civil rights out of the "single-issue ghetto" and brilliantly exploited the common ground between civil, social, and economic justice, pointing to underlying linkages among racism, poverty, and militarism. Today's environmental justice leaders are doing essentially the same thing—removing environment from its single-issue ghetto and linking it with civil and human rights.

When King realized that his alleged allies in Washington could do the movement more harm than the segregationists—by trading away basic civil rights for short-term gains—he escalated direct actions at the local level, often in defiance of pleas for moderation from Washington allies. When he persisted with nonviolent confrontations they responded to his actions and rhetoric very much as today's mainstream environmental leaders respond to grassroots activists. Thurgood Marshall believed that the only way to establish civil rights was through litigation and that King's protest marches were counterproductive. "King is a rabble-rouser . . . a boy on a man's mission . . . an opportunist," bellowed Marshall as he prepared his historic brief for the Supreme Court. "All that walking for nothing," he exclaimed after winning Brown *vs.* Board of Education. "They might as well have waited for the Court's decision." The same sort of naiveté has, unfortunately, infected many environmental leaders, who seem confident that the real victories will be won in court and Congress without rabble-rousers in the streets, the woods, and the factories.

King succeeded because he ignored Thurgood Marshall's admonitions and imposed his civil rights agenda on every institution in the land. Until environmentalism is similarly institutionalized—into academia, nonenvironmental government agencies, and religious institutions—it is unlikely that major new gains will be made. "King saw three major evils," writes environmental historian Michael Frome, "racism, poverty and militarism—and found them to be integrally linked, one with the other. I see the environment as a fourth major evil, also joined with the others." King, had he lived a few more years, would almost certainly have come to the same conclusion and agreed with Frome that "environmentalists speak of concern with forests, water, air, soil, fish and wildlife, land use and the use of resources, but these are only symptoms of a sick society that needs to deal more fundamentally with itself."[43]

Just as the civil rights movement called to task the basic orientation of all unjust societies and institutions, environmentalists—and not just radicals and greens—need to challenge the basic orientation of all industrial economies (capitalist, socialist, and communist), even if it means losing corporate and conservative foundation support. If the past 20 years of environmentalism have taught us anything it is that when the threat of public governance of productive technologies disappears, polluters relax. Regulatory flexibility and "constructive engagement" with industry have created some green business heroes, but they can be counted on one hand. The rest, unfortunately need to be regulated.

Were King still alive he would have by now confronted the mainstream environmental movement with its inherent racism. He would also have attempted to create détente between the mainstream and the grassroots of the movement, encouraging joint projects between Green Group organizations and grassroots coalitions. King would have seen Lois Gibbs' organizing skills and Fred Krupp's access to power as ingredients of a powerful alliance. Without such alliances the grassroots movement will almost certainly become the leading edge of America's environmental future, simply because their members are the most active, passionate, committed, and angry. The comparatively passive supporters of mainstream groups have proven themselves mercurial, faddish, and easily attracted to other causes.

Were mainstream leaders to overcome their cultural distrust of direct action and activate their members at the local level things would change rapidly. Organized boycotts, staged demonstrations, shareholder suits, and, even, using some of the tactics that made Greenpeace the largest environmental organization in the world could motivate members to remain loyal and polluters to pay more heed. A thousand hunters and fishermen waving their NWF membership cards at the gates of DuPont would have far greater impact than 10 times as many students in Birkenstocks.

Conclusion

On a practical level conservation has been sustained by an interplay between professionals and radical amateurs. Professionals keep the movement organized. Amateurs keep it honest. The ghosts of Muir and Pinchot still wrestle for control—in a fractious but symbiotic embrace.

Stephen Fox, John Muir and His Legacy

Agitate, litigate, legislate!

Martin Luthor King, Jr.

As conservation consciousness progressed toward environmental consciousness, humans, once considered an elite life form, the crown of evolution manufactured in the image of their maker, began to regard themselves as merely another species—seen by some as an endangered species, by others as a dangerous species.

Although some ideologies accord other living things equal status with humans, American environmental laws and mandates are still largely oriented around protecting people and their health. Of all the federal laws passed during the era of environmental legislation, only the Endangered Species Act protects nonhuman life, and even that with a prohuman caveat. Yet fourth-wave thinkers, like indigenous peoples around the world, believe that to protect human life, all life must be protected; to protect human health, all life must be healthy. Such activists will not, therefore, restrict environmentalism to the protection of human health, a narrow concept that has done so little to protect land, air, water, *or* human health. Fourth-wave thinking and activism will protect ecosystems, biodiversity, habitats, and aquifers. Human health will follow naturally from such a course.

A redefined environmentalism will come gradually through a synthesis of classical environmentalism and the emerging ecologies. The fourth-wave environmental ideology will lean cautiously toward the human-in-nature world view and blend primitive earth wisdom with the verities of modern scientific ecology. It will respectfully reject the anthropocentrism of humanist, Marxist, and Judaeo-Christian traditions, challenge the Newtonian-Cartesian view and call for a new philosophy of nature—a new ecological ethos that says: "We are aware now of our impact on the environment and the planet. The next step is to see ourselves as citizens of nature." Or, to paraphrase Thomas Berry, "Human beings will reapply for membership in the biosphere."

The so-called new ecologies have been regarded with suspicion only partly because they were "primitive," "pagan," or "radical," or because they were too quickly embraced by monkeywrenchers, eco-terrorists, and back-to-the-land hippies. The ecologies also challenge many of the fundamental premises of prevailing economic orders—not only market capitalism, but *any* corporate or bureaucratically dominated, heavy industrial, consumer-driven economy dependent on mass centralized technologies—capitalist, communist, or bureaucratic socialist.

The founders and early leaders of today's mainstream organizations brought to their work a passionate commitment to the cause and a fairly clear understanding of their relationship to nature. They read their Muir and their Leopold. What they lacked in organizational skills, they made up in zeal. There are still people with such sensibilities in the mainstream movement, but they are not in the top jobs. Many of them leave Washington and New York every year to work in smaller vineyards. When they leave, the bureaucracies they abandon become a little colder.

Without fervor in the ranks the nationals cannot formulate a vision. Broaching questions about the future with almost any mainstream leader will draw one into a discussion of federal rulemaking, organizational development, or fundraising strategies—likely items on the agenda of the next board meeting. When leaders do take time to reconsider their mission and explore a vision of the future (as happened recently at the Natural Resources Defense Council), they find themselves restrained by the imperatives of their benefactors and the sheer size of their organizations. They tend to continue on the track that has kept them alive thus far, whether it protects the environment or not. As my friend Al Meyerhoff told me recently on a plane to New York, NRDC has completed its review of mission and intends to continue litigating. It was beginning to pay off, he said, because the Clinton administration had replaced a few troglodyte Reagan appointees with better judges.

Less than a week later Meyerhoff and a group of plaintiffs reached a settlement with the EPA that could end the use of 36 pesticides found to be carcinogenic in animals. Of course it is an open question whether the EPA settled because they feared losing in court or because Carol Browner and the administration had taken so much heat from grassroots environmentalists for threatening to weaken the Delaney clause, which prohibits all carcinogens from processed foods. It's a matter of perspective.

Rising through the ranks of NRDC and the other nationals are scores of young people who embrace the new ecologies, or aspects of them. They do so quietly so as not to disturb their leaders. They can be found at Buddhist retreats and vision quests exploring the spiritual side of ecology and their lives. In those paneled Washington offices there are also radicals who, like David Brower and Dave Foreman before them, are losing patience with slow pragmatic incrementalism and endless compromise. Although some of them may become leaders of the mainstream, many others will leave, disillusioned, and join forces with the grassroots. On both sides of that unfortunate divide between reform and populist movements they will forge a synthesis of vision and pragmatism that will produce the fourth wave of American environmentalism.

Epilogue

A Reason (or Two) for Hope

Like many others, I believe that our present form of technological civilization will eventually collapse if present trends continue—but what a big *if* this is.
René Dubos

After reading this book you may be surprised to hear that I am optimistic about the future of environmentalism in America. Like Dubos, I must at times describe myself as a "despairing optimist"—particularly on days when I hear that the frog population of the world is down and that algae blooms are up; that ozone holes are opening again somewhere over the planet or that fish counts are low; that song birds are disappearing, deep aquifers are drying up, or a cluster of human birth defects has appeared in some town held in high esteem for its industrious spirit.[1] Despite such foreboding signals, which I seem to hear more often than I used to, I remain an optimist. Here is why.

American civilization developed a vivid environmental imagination, then became the first society in history to turn that imagination into a political movement. The educational arm of the movement has taught Americans a very simple but very important principle (and most understand it). It is that we all exist within an ecological construct called "the environment" and that we allow compromise of the environment at our own peril. The environmental movement is in crisis, as this book attests, but it is very much alive. And far from its nucleus, on the outer edges of its vast penumbra, there are signs of rejuvenation.

Every poll that asks about it testifies that environmentalism is deeply rooted in the American psyche. Despite assault from prevalent ideologies—capitalism, materialism, apocalyptic religion—and in defiance of powerful institutions and antagonistic counter-movements, environmentalism persists. Skeptics insist that it can always be submerged by baser instincts. That may be true for a while but not for long, for the American environmental imagination is deep, and it is hard-wired to the American dream.

The water flowing from the tap in that imaginary three-bedroom house is pure and potable. The chicken in the pot is range fed and salmonella-free. The car in the garage burns clean and runs a long way on a gallon of unleaded gas or, better still, propane. The fruit and vegetables in the shopping bag are nutritious and pesticide-free, and there is a wide meadow somewhere in the mountains where the family can pitch a tent and spend a few days in the wild. New ideologies and deeper ecologies will come and go, but in the end it is imagination that drives this movement.

During the summer of 1988 a series of startling images provoked that imagination. Drought ravaged the south and west of the country. Crops and livestock perished in the fields. In the midst of an unbearable heat wave a large flotilla of used hypodermic needles washed up along the Atlantic shore of North America. Beaches were closed. Across the ocean thousands of dead seals were found floating in the North Sea. Eight thousand miles south, sheep were going blind in Tierra del Fuego. Yellowstone Park caught fire. Through it all, the daily television news tracked a vagrant garbage barge called *Mobro* as it was towed from port to port in search of a dump in which to unload its stinking refuse. *Time* magazine named Earth "planet of the year." "This year the earth spoke, like God warning Noah of the deluge," wrote the editors (failing to capitalize the name of their honored celebrity). "Its message was loud and clear, and suddenly people began to listen, to ponder what portents the message held."[2]

It would have been easy to despair. But the eco-episodes of the summer of '88 woke America from the long somnambulance of the Reagan era. Sometime toward the end of that portentous year, the 1990s were declared "the Decade of Decision" for the American environment. "Not because we say so," cautioned eco-prophet, Fritjof Capra, "but because of events almost beyond our control."[3]

That declaration made the 1990s the third proclaimed "environmental decade" in a row, raising the question of how many environmental decades would be required to save the Earth. The American answer appears to be "as many as it takes." The probable answer is an infinite number, for *homo industrius* seems determined to challenge nature's (or is it Gaia's?) ability to counterbalance humanity's environmental assaults; in which case a perpetual movement is required.

Although the most visible leaders and institutions of the movement remained indecisive and irresolute, the early 1990s saw the stirrings of vital new energies at the grassroots, where literally thousands of tiny ad hoc groups often derided as "NIMBYs" were formed. Run by people who, 10 years ago, would never have called themselves "environmentalists," they began fighting separately to save their lives and together to "plug the toilet" of industrial pollution. They give me hope.

In the towns and villages surrounded by our national forests, and populated by families dependent on logging and milling, hundreds of citizens have formed organizations to stop the clearcutting of the last few ancient trees left on the continent. They withstand the anger of loggers whipped up by junk-bond corporados and Wise Use vigilantes. They face the abject corruption of the U.S. Forest Service and shake their heads over the persistent equivocations of mainstream leaders. Yet these groups, this movement-within-a-movement, fighting big industry one timber sale at a time, has saved more trees in the past 10 years than all the legislation passed in Washington. They give me a reason for hope.

The mainstream leaders I interviewed often belittled these grassroots uprisings as parochial, pointing to their obvious lack of professionalism, inability to reach the national media, and single-issue obsessions. "They're utopian . . . just blowing smoke," wrote Sierra Club Director Michael McCloskey to his board in January 1986.[4] One national leader I interviewed dismissed the entire struggle for environmental justice as "opportunistic" and "temporary" (not for attribution of course). Mainstreamers still seem confident that with enough money and professional training they can prevail with conventional weapons.

Grassroots activists, on the other hand, seem to sense that much more is riding on the ability of the American environmental movement to reinvent itself than the success of a legislative strategy, or the ability to prod scofflaw regulators with litigation. "With the life-sustaining capability of the biosphere at stake, this is no time for tired old tactics" one activist told me, "nor are incrementalist strategies adequate." What is clearly surfacing within the American environmental movement is some much-needed belly fire and a willingness to be audacious, confrontational, unpopular, and unphotogenic. That gives me hope.

My hope also derives from the disparate reaches of the movement where respect for ideological diversity is growing alongside a struggle for justice. There may be no single ideology that can unite the diverse interests of American environmentalists, they seem to be saying, only the single goal of justice. That's all right. Many environmentalisms and many ideologies fed into this movement. There is no reason why they can't synthesize slowly. Diversity can be preserved and tolerated among wilderness preservationists, anti-toxic campaigners, animal rights advocates, anti-nuclear activists, feminist ecologists, green politicos, population-control zealots, and neo-Luddites—if they each perceive and act upon the essential injustice of environmental degradation.

There is hope too when those for whom human welfare is the highest priority begin to accept Aldo Leopold's unwitting Buddhist maxim that "all ethics rest upon this single premise: That the individual is a member of interdependent parts." Or when human rights advocates understand Philip Shabecoff's axiom that "*all* nature has a right to exist for itself." Hope dawns again when Dana Alston assures the People of Color Summit that they are convened not in reaction to the racial majority that dominates environmentalism but to "affirm our traditional connections to and respect for the natural world." That convention went on to formulate a new set of principles for environmentalism (see appendix), which begins: "Environmental justice affirms the sacredness of Mother Earth, ecological unity, the interdependence of all species, and the right to be free from ecological destruction." There is hope too in the sentiment of John Adams, who speaks passionately about the plight of people of color, and in the speeches of Robert Bullard, newly salted with images of endangered whales and butterflies.

With prevailing economic systems in various degrees of disarray, the worldwide environmental movement seems uniquely positioned to serve as a vehicle for a civilization ordered on a new basis. It is clear now that both capitalism and socialism, in all their experimental forms, have failed to create ecologically sustainable economies. Environmentalism, with its broad multi-partisan appeal, permits political leaders and economic visionaries to conceive a third way, between the market and the plan, and to do so in a context that carries considerable moral weight. It is heartening then to see the environment being injected into the previously forbidden territory of debates over trade and international relations.

The movement we now support may fail. But environmentalism itself will survive. There is something very enduring about this nation's imagination. But if American environmentalism is to become "the single most important movement of the twentieth century"—as historians and social scientists have predicted—the movement needs to be as large as the environment itself. Environmentalism needs to penetrate every institution, ideology, and religious faith in our culture. It needs to be seen as a social as well as a political movement. That appears to be happening, perhaps because in our deepest consciousness we know that without it we will perish and that whatever other priorities we temporarily place before it will also perish.

I began this project believing that the only hope for the environment was its rediscovery by a whole new movement led by the twenty-first-century descendants of Henry David Thoreau, John Muir, Aldo Leopold, Edward Abbey, Alice Hamilton, Bob Marshall, Rachel Carson, and Martin Luther King. I still believe that. And I believe it is happening. The American environmental movement has only begun.

Notes

Epigraph

1. In John Berger's novel *Pig Earth* (London: Writers and Readers Publishing Cooperative, 1979) the son of a farmer in southern France returns from a summer job in Paris, where he had seen, in the lobby of a bank, an old wine press from his home village. His father, baffled that a useful tool had been turned into a piece of sculpture, grumbled "the world has left the Earth behind it."

Preface

1. Quoted by Lynton Caldwell, "Globalizing Environmentalism: Thresholds of a New Phase in International Relations," *Society and Natural Resources*, 4, p. 259.

2. Walinsky, when he wrote this for *Earth Day—The Beginning: A Guide For Survival* (New York: Arno Press), a collection of essays compiled by Environmental Action, was a Democratic candidate for attorney general of New York and a former advisor to the late Robert F. Kennedy.

3. Riley E. Dunlap, "Trends in Public Opinion Toward Environmental Issues," *Society and Natural Resources*, 4, p. 286.

Introduction

1. William T. Hornaday, *Wildlife Conservation in Theory and Practice, Lectures Delivered before the Forest School of Yale University* (New Haven: Yale University Press, 1914), p. 191; Madison Grant, *The Passing of The Great Race* (New York: Scribners, 1916), pp. 14, 23, 46, 73, 81.

2. Eventually the environmental movement itself was branded a communist plot by the John Birch Society, which suspected that Earth Day 1970 was a covert attempt to celebrate Lenin's birthday—which fell on April 22 (It is also the birthday of Francis of Assisi and Spanish Queen Isabella!).

3. *The New York Times* poll, 1992.

4. Barry Commoner estimates in his book *Making Peace With the Planet* that about $1 trillion in public and private money has been devoted to the environment in the United States.

5. In 1990 the National Audubon Society had 578,000 members; by 1993 it had dropped to 542,000. In the same period the Sierra Club dropped from 622,000 to 550,000, the Wilderness Society from 440,000 to 300,000 and Greenpeace from 2.5 million to 1.7 million.

6. Prediction made by Robert Nisbet, quoted by Lynton Caldwell in "Globalizing Environmentalism," *Society and Natural Resources*, vol. 4, p. 259.

Chapter 1

1. Historian Donald Worster claims that "no other people in the world has ever believed, as Americans have, that they are actually living in Eden." *The Wealth of Nature: Environmental History and the Ecological Imagination* (New York: Oxford University Press, 1993), p. 9.

2. Genesis 1:28. Some translations of the text read "fill the earth and subdue it." The King James version reads "replenish the earth and subdue it."

3. Jean-Jacques Rousseau, *Discourse on the Origins and Foundations of Inequality Among Men* (1754).

4. All quoted in Keith Thomas, *Man and the Natural World* (New York: Pantheon Books, 1983).

5. Quoted in Philip Shabecoff, *A Fierce Green Fire: The American Environmental Movement* (New York: Hill and Wang, 1993), p. 54.

6. The name *Gaia* should actually be attributed to British novelist William Golding, a neighbor of Lovelock's in Devon, who, upon hearing Lovelock's hypothesis, suggested naming it for the Greek earth goddess.

7. Bison hunters occasionally took the hides of a few of the animals killed, and there was a small market for the tongue. Most often the kill was left to rot in a huge pile in the sun.

8. Hunters and fishermen still hold a strong sway over policy in many mainstream organizations. Members of the National Wildlife Federation and the Izaak Walton League are largely "hook and bullet" types. NWF leaders are wary of participating in coalitions with Defenders of Wildlife, a smaller organization that opposes most hunting on national reserves. As recently as 1994, the National Audubon Society, which has struggled over the issues of hunting and trapping on public lands since its origins, found itself deferring to the hunters in its membership over the content of an Audubon-produced television documentary about the U.S. Wildlife Refuge system. Scenes of hunting and trapping that were portrayed as questionable "secondary uses" of the reserves were removed from the film on orders from Society President Peter A. A. Berle.

9. Gifford Pinchot, *Breaking New Ground* (New York: Harcourt Brace, 1947), p. 23.

10. Quoted in Roderick Nash, *Wilderness and the American Mind* (New Haven: Yale University Press, 1967), pp. 179–80.

11. Robert Gottlieb, *Forcing the Spring: The Transformation of the American Environmental Movement* (Washington, D.C.: Island Press, 1993), p. 19.

12. Aldo Leopold, *A Sand County Almanac* (New York: Oxford University Press, 1949, 1953).

13. Ibid.

14. Curt Meine, *Aldo Leopold, His Life and Work* (Madison: University of Wisconsin Press, 1988), p. 363.

15. Worster, *The Wealth of Nature,* pp. 104–105.

16. Aesthetics often muddle the environmental mind. As P. R. Hay points out, "from an ecological perspective, some of the areas most meriting protection have no value as scenery at all."

17. *Silent Spring* (Boston: Houghton Mifflin, 1962). Robert Downs lists *Silent Spring* high among "books that changed America," and John Kenneth Galbraith calls it one of the most important books of western literature. Stuart Udall, Secretary of Interior under Kennedy and Johnson, dubbed Carson "the fountainhead" of the new environmental movement. And historian Stephen Fox sees *Silent Spring* as "the Uncle Tom's Cabin of modern environmentalism."

18. Daniel Kevles, "Greens in America," *New York Review of Books,* October 6, 1994, p. 36.

19. Robert Gottlieb, *Forcing the Spring,* p. 47.

20. Riley Dunlap and Angela Mertig, *American Environmentalism: The U.S. Environmental Movement, 1970–1990* (Philadelphia: Taylor and Francis, 1992).

21. Riley Dunlap, Angela Mertig, and Robert Cameron Mitchell, "Twenty Years of Environmental Mobilization," *Society and Natural Resources,* vol. 4, p. 222.

22. Samuel P. Hays, "From Conservation to Environment: Environmental Politics in the United States Since World War Two," *Environmental History Review,* vol. 6, no. 2 (Fall 1982), pp. 14–41.

23. Quoted in *PR Watch,* second quarter 1994.

24. Remark made to an AFL-CIO meeting shortly after announcing "A National Teach-in on the Crisis of the Environment." Quoted in Gottlieb, *Forcing the Spring,* p. 106.

25. Edgar Wayburn, "Survival Is Not Enough," *Sierra Club Bulletin,* March 1970, p. 2.

26. Quoted in Kirkpatrick Sale, *Green Revolution: The American Environmental Movement, 1962–1992* (New York: Hill and Wang, 1993).

Chapter 2

1. In a memorandum written January 19, 1986, to the Sierra Club board McCloskey warned against the "new, more militant" environmentalists who "would not hesitate to criticize the main players such as the Sierra Club. . . . They're just utopian."

2. Elaine Koerner *Science Rules; Priority Setting Within Environmental Organizations* (Washington D.C.: EPA Office of Public Liaison). The full list, ranked in order of priority, was ecosystems, environmental education, laws and legislation, water, air, global issues, hazardous waste, pollution prevention, solid waste, toxics, human health, research and technology.

3. Shabecoff, Philip, *A Fierce Green Fire* (New York: Hill and Wang, 1993), p. 112.

4. One of the young Yalies was John Bryson, who moved west to open the San Francisco office of NRDC. Soon after the office was opened he was appointed by Governor Jerry Brown to the State Water Resources Control Board and then to the Public Utilities Commission. From there he was recruited by Southern California Edison, the state's largest public utility. He is today its CEO. Ironically, Bryson now finds himself embroiled in a major clean air controversy over construction of a massive coal-fueled electric plant California Edison built in Mexico to Mexican air pollution standards. The plant stands across the Rio Grande River from Big Bend National Park in Texas, which will be directly downwind from the plant. Bryson claims that Southern California Edison cannot afford to bring the plant up to American emission standards.

5. At the time, the Sierra Club president was Laurence Moss, a nuclear engineer who discouraged club opposition to nuclear energy.

6. Thomas Turner, "The Legal Eagles," *Crossroads* (Washington D.C.: Island Press), p. 52.

7. Robert Gottlieb, *Forcing the Spring: The Transformation of the American Environmental Movement* (Washington D.C.: Island Press, 1993), p. 139.

8. Interview with Meyerhoff.

9. Ibid.

10. Barry Commoner, "The Failure of the Environmental Effort," *Current History* (April 1992), pp. 176, 177.

11. Ibid. Commoner ran for president as a candidate of the Citizens Party in 1980.

12. According to the Foundation Center, about half of all American philanthropy goes to religious institutions. Individuals contribute 81 percent; foundations, 7 percent; bequests, 7 percent; and corporations, 5 percent. Dead people contribute more than corporations. During the years that profits rose 100 percent, the latter increased their contributions by only 26 percent. The dollar figures given in this paragraph are not adjusted for inflation.

13. Nathan Newman and Anders Schneiderman, "Market Leftism," *Crossroads* (May 1994), p. 17.

14. Interview with Gillenkirk.

15. *National Environmental Study: Report and Supporting Data,* EOS (1993).

16. Environmental monthlies rely in varying degrees on revenue from national advertisers, who demand coated stock to convey their message. Thus they have been reluctant to print their magazines on anything less that 20 percent post-consumer recycled paper. *Sierra* was the first to change to recycled paper, in November 1990.

17. Craver, Matthews and Smith, direct mail consultants in Washington D.C. A substantial portion of the rest, particularly at the National Wildlife Foundation, comes from the sale of posters, tee shirts, coffee mugs, and magazines.

18. EOS poll survey.

19. Interview with Schifferle.

20. Many believe that Greenpeace lost the bulk of its members during this period because of its vocal oppositon to the Gulf War.

21. Inteview with Sher.

22. Interview with Gillenkirk.

23. Pew Charitable Trust, *Environmental Strategies: Concept Statement* (December, 1993), p. 3.

24. Ibid., p. 4.

25. Ibid., p. 6.

26. Interview on September 26, 1994, with Josh Reichert, Environmental Director, Pew Charitable Trust.

27. Interview with Tim Hermack.

28. Green, Stephen G., "Who's Driving the Environmental Movement?" *The Chronicle of Philanthropy* (January 25, 1994), p. 6.

29. Quoted in *Chronicles of Philanthropy,* November 17, 1992, p. 7.

30. Quoted in William Greider, *Who Will Tell The People, The Betrayal of American Democracy* (New York: Simon and Schuster, 1992).

31. "Links With Activists Get Results in Environmental PR," *O'Dwyer's PR Services,* vol. 8, no. 2 (Feb. 1994), p. 1.

32. Quoted in "Corporations Jumping on the Environmental Bandwagon by Giving Dollars and Time," *Corporate Giving Watch,* vol. 12, no. 3 (May 1993), p. 1.

33. Quoted in Greider, *Who Will Tell the People,* p. 24.

34. *Corporate Philanthropy Report,* vol. 7, no. 8 (May 1992), p. 1.

35. Interview with Ayres.

36. *O'Dwyer's PR Services,*, Feb. 1994.

37. Ibid. (Italics in original.)

38. Waste Management, Inc. also became a matter of great controversy within the Environmental Grantmakers Association (EGA). Formed in 1987 by a handful of large foundations, EGA is a loose-knit organization of environmental philanthropists, mostly foundations, designed to exchange ideas, discuss strategy, and persuade other philanthropists to support environmental activity. From the beginning, the question of who could join EGA and who could attend meetings has been a source of heated disagreement. The first test of tolerance came in 1988 when Waste Management, Inc. attempted to join EGA. It was immediately rebuffed. An internal memo reads: "It is readily apparent that Waste Management has engaged in a pattern of abusive corporate conduct involving repeated violations of both criminal and civil laws." Following heated debate EGA adopted the following policy: "EGA membership is open to any foundation or corporate giving program that chooses to join, with the sole exception of one that during the past five years has been convicted of felony offense directly related to environmental degradation or whose employee in the course of employment or whose majority-owned subsidiary has been convicted of such an offense."

The absence of any concept of social justice in the policy horrified grassroots environmentalists. The loose legalistic screen for EGA membership took account of a corporation's criminal offenses but completely ignored its civil offenses. The idea that a company had to be convicted of a felonious environmental offense to be considered environmentally unworthy simply made no sense to them.

39. Jim Donahue, "Environmental Board Games," *Multinational Monitor* (March 1990), p. 10. The boards surveyed were those of the Environmental Defense Fund, the National Audubon Society, the National Wildlife Federation, the Natural Resources Defense Council, the Wilderness Society, the World Resources Institute, and the World Wildlife Fund.

40. Quoted in Ibid., p. 11.

41. Statement of Peter A. A. Berle, reprinted in *The Workbook,* vol. 18, no. 1 (Spring 1993), p. 7.

42. Ibid.

43. Richard Regan, "How Much Influence Does Corporate Money Buy?" *The Workbook,* vol. 18, no. 1 (Spring 1993), p. 11. Not everyone in the environmental movement accepts WMI funds. In October 1991 the Lutheran Church in Germany held a conference of environmentalists and corporate leaders to discuss environmental responsibility. When leaders of the Citizens Clearing House on Hazardous Wastes and the National Toxics campaign discovered that WMI was a funder of the conference they politely declined the invitation, at which point the organizers of the conference politely told WMI to keep its money.

44. Interview with Newsom.

45. Amyas Ames, "Drafted into the Movement," *Acorn Days,* p. 63.

46. Sale, *Green Revolution: The American Environmental Movement, 1962–1992* (New York: Hill and Wang, 1993), p. 56.

Chapter 3

1. George Wald, "Environmental Traps," in *Earth Day—The Beginning, A Guide to Survival* (New York: Arno Press. 1970), p. 80.

2. Adlai Steveson III, "Too Little Too Late," in *Earth Day—The Beginning,* p. 53.

3. The Reagan administration could not have been more wrong in its reading of public attitudes toward environmentalism. In April and September of 1981 the Harris Poll found that more than 80 percent of the public favored maintaining the Clean Air Act or making it stricter.

4. Between 1982 and 1985 air particulates monitored by other agencies nationwide rose 4 percent a year. Sulfur dioxide emissions, which declined dramatically from 1975 to 1981, remained essentially constant throughout the Reagan presidency. Carbon monoxide, which had also decreased dramatically between 1975 and 1981, began to rise in 1982.

5. William Reilly, director of the Conservation Foundation, declined the invitation, expressing concern about aligning his organization with adversarial groups like EDF and NRDC.

6. Interview with Robert Allen.

7. Ibid.

8. Interview with Bill Turnage.

9. Allen, ibid.; Michael Fischer, Memorandum to John Adams, NRDC, April 25, 1994.

10. Robert Mitchell, Angela Mertig, and Riley Dunlap. "Twenty Years of Environmental Mobilization," *Society and Natural Resources,* vol. 4, p. 223.

11. Interview with Richard Ayres.

12. Ed Marston, *High Country News,* Feb. 3, 1986.

13. Riley Dunlap, "Trends in Public Opinion Toward Environmental Issues," *Society and Natural Resources,* vol. 4, p. 299.

14. Interview with Peter Bahouth.

15. Fischer, Memorandum, April 25, 1994.

16. Ibid.

17. David R. Brower, *For Earth's Sake: The Life and Times of David Brower* (Salt Lake City: Peregrine Smith Books, 1990), p. 350.

18. "The Government Is the Environment's Worst Enemy," Earth Island Journal (Fall 1991), p. 39. Rick Sutherland died tragically in an auto accident shortly after he wrote this. Many believe he was one of the few lawyers who realized that the legislative/litigative strategy was failing and had the imagination and leadership to fix it. He was succeeded at SCLDF by Vic Sher.

19. William Greider, *Who Will Tell The People: The Betrayal of American Democracy* (New York: Simon and Schuster, 1992), p. 213.

20. Former District of Columbia Appeals Judge Robert Bork, writing in the *Wall Street Journal* on October 13, 1989, expressed the sentiments of many on the federal bench: "These last two decades it has come to be thought that individuals can go to court to assert their own parochial views of the public's legal rights. This is contrary to the traditional rule that a citizen cannot sue a prosecutor to require him to enforce law in a particular way or even to enforce the law at all."

21. Keith Schneider. *The New York Times,* March 23, 1992, p. A1.

22. Ibid.

23. Greider, *Who Will Tell the People*, p. 149.

24. *The New York Times,* March 23, 1992, p. A1.

25. Quoted in Greider, *Who Will Tell the People*, p. 171.

26. Ibid.

Chapter 4

1. Quoted in Frank Graham, *Since Silent Spring,* (Boston: Houghton Mifflin, 1970), p. 49.

2. *Silent Spring* stimulated the creation of a Presidential Scientific Advisory Committee that, in 1963, corroborated Carson's findings. The report became the impetus for early drafts of both the Pesticide Control Act of 1972 and the Toxics Substances Control Act of 1976.

3. *Total PAC and Individual Contributions in Congressional Campaigns 1991-92* (Washington D.C.: Center for Responsive Politics, 1993).

4. *Sierra* Magazine, October 1994, p. 28.

5. Quoted in William Greider, *Who Will Tell the People: The Betrayal of American Democracy* (New York: Simon and Schuster, 1992), p. 216.

6. Eric Mann, "Environmentalism in the Corporate Climate," *Tikkun*, vol. 5, no. 2 (February 1990).

7. Donald Snow, "Wise Use and Public Lands in the West," *Northern Lights* (Winter 1994).

8. Quoted in David Helvarg, *The War Against the Greens* (San Francisco: Sierra Books, 1994), p. 137. In the same interview with Helvarg, Gottlieb added: "What's really good about it is it touches the same kind of anger as the gun stuff, and not only generates a higher rate of return but also a higher average dollar donation. My gun stuff runs about $18. The Wise Use stuff breaks $40." "Pay out" is the rate at which a direct mail entrepreneur like Gottlieb gets his own money back.

9. Ibid.

10. Ibid., p. 134.

11. Ibid., p. 124.

12. Michael O'Keeffe and Kevin Daley, "Checking the Right," *Buzzworm, The Environmental Journal,* vol. 5, no. 3 (May/June 1993), p. 44.

13. Rush N. Limbaugh, *The Way Things Ought to Be* (New York: Pocket Books, 1992), p. 116.

14. O'Keefe and Daley, "Checking the Right," p. 42.

15. Helvarg, *War Against the Greens.*

16. Eve Pell, "Stop the Greens," *E: The Environmental Magazine* (Nov./Dec. 1991), p. 34.

17. Helvarg, *War Against the Greens.*

18. Watt's remark was made at a Cattleman's Dinner in Denver, Colorado, June 1990; information on western U.S. vigilantism in Helvarg, *War Against the Greens.*

19. The order was inspired by Richard A. Epstein's book, *Takings, Private Property and the Power of Eminent Domain* (Cambridge: Harvard University Press, 1985).

20. Helvarg, *War Against the Greens,* p. 22.

21. Thomas M. Power, "The Price of Everything," *Sierra* (Nov./Dec. 1993), p. 92.

22. *New York Times,* January 20, 1992.

23. Interview with Willer.

24. Interview with Reichert.

25. Interview with Kerr.

Chapter 5

1. *The Mandate for Leadership* (1980) advocates "market mechanisms to induce individual firms to find the most efficient means to stay within pollution limits." Hair's remark is quoted from the *Washington Post,* August 4, 1991.

2. Sherri Bittenbender and Robert Stavins, Project 88: Harnessing Market Forces to Protect Our Environment, Harvard Energy and Environmental Policy Center working paper, 1989.

3. Quoted in Craig Mellow, "The McGreening of America," *Inc.* (February 1991), p. 62.

4. William Reilly, *The Greening of EPA,* EPA, 1989.

5. From EDF promotional material.

6. Quoted in Jonathan Dushoff, "A License to Pollute," *Multinational Monitor,* March 1990, p. 18.

7. On March 31, 1993, the Chicago Board of Trade (CBT) held its first auction of pollution credits, announcing the opening of bids competing for 150,000 allowances. The 171 sealed bids came not only from electric utilities seeking to trade their "rights to pollute" with other utilities, but also from brokerage houses and private investors anxious to speculate in the pollution credit market. The first day's trading alone brought $21 million onto the CBT trading floor. *See* Jeffrey Lewis, *Wall Street Journal,* August 24, 1993, p. C1.

8. Several aggressive brokerage firms have since established an over-the-counter market for pollution credits. "We don't feel the Board of Trade is going to fit the needs of the industry," said John B. Henry, president of Clean Air Capital Markets. (*Wall Street Journal,* August 24, 1993).

One positive use of emissions credits was made by the owner of 10,000 tons of sulphur dioxide credits worth $3 million on the open market. The company donated them to the American Lung Association in New England, which was able to "retire" that amount of SO_2 from the region's air. Working Assets Company in San Francisco has also purchased pollution credits with no intention of selling them or exercising the right to pollute.

9. The Clean Air Act of 1990 also allows utilities governed by state regulations to trade in the federal pollution credit market. One of the first ERC trades made was the sale of 20,000 tons of SO_2 credits by Wisconsin Electric Power to a Pennsylvania utility. Under Wisconsin law WEP was not entitled to emit that tonnage in the first place, yet the federal law allowed it to sell the right to Pennsylvania.

10. Daniel A. Seligman, "Air Pollution Emissions Trading: Opportunity or Scam?" Sierra Club White Paper.

11. Ibid.

12. One intriguing marketing scheme is called Global Warming Alternatives (GWA), a California firm that touts iself as "the first company in the U.S. to commercialize carbon offsetting as a market-based solution to global warming." GWA offers a portfolio of "offsetting" investments to heavy emitters of greenhouse gases. The portfolio purportedly contains "carbon dioxide offsets"—investments in "energy efficiency, conservation, and

biomass-based carbon sequestration." Shares in the portfolio will be sold to heavy CO_2 polluters such as public utilities, as well as to "socially responsible corporations, green consumers, the oil industry, the automotive industry and green states and foreign governments."

13. John Lancaster, "The Environmentalist As Insider," *Washington Post* Magazine, August 4, 1981, pp. 17–29. In 1988 a number of National Wildlife Federation members complained directly to Jay Hair about the presence of Dean Buntrock on the NWF Board. Hair's response, contained in a February 1988 form letter, read: "We feel that Waste Management, Inc. is conducting itself in a responsible manner." By then Waste Management had already broken most records for toxic-handling violations and federal fines.

14. WMX alone has contributed more than $1 million to environmental groups, which may explain why Buntrock sits on the board of the National Wildlife Federation. The Conservation Foundation received generous donations from Chevron, Exxon, General Electric, Union Carbide, Weyerhaeuser, WMX, and other corporations the year before its president, William Reilly, became EPA administrator.

15. Rose Gutfeld, *Wall Street Journal,* August 20, 1992, p. B1

16. Caption under cartoon of a North American business tycoon shouting from his limousine to a South American peasant about to chop down a tree in the rain forest.

17. The term *ecological imperialism* was used in a resolution passed on May 10, 1991, by 21 Ecuadorian environmental and indigenous peoples' organizations meeting in Puyo, Ecuador. It is often used by the third world organizations and others to describe the activities of American environmental envoys.

18. Letter signed by 13 Ecuadorian organizations, May 25, 1991, Quito, Ecuador.

19. Rachael Carson used this quotation as the dedication to *Silent Spring*.

20. Ann Reilly Dowd, *Fortune,* September 19, 1994, p. 104.

21. Interview with Ayres.

22. Thomas M. Power and Paul Rauber, "The Price of Everything," *Sierra,* Nov./Dec. 1993.

23. Ibid.

Chapter 6

1. Dorceta Taylor, *Proceedings of the Michigan Conference on Race and the Incidence of Environmental Hazards* (July, 1990).

2. Andrew Szasz, *Ecopopulism: Toxic Waste and the Movement for Environmental Justice* (Minneapolis: University of Minnesota Press, 1994), p. 40.

3. Ibid., p. 42.

4. Robert Gottlieb, *Forcing The Spring: The Transformation of the American Environmental Movement,* (Washington, D.C.: Island Press, 1993), p. 116.

5. Hickman and Brunner are quoted in Barry Commoner, *Making Peace With the Planet* (New York: Pantheon, 1990), p. 121.

6. O'Connor called the dissolution "a reorganization" and assembled a new organization called the Jobs and Environment Campaign. Gary Cohen, NTCF's top fundraiser, also sees the dissolution a positive step. "When organizations that are set up to serve people can no longer do so they need to be swept aside so that other, more effective organizations can take their place." When NTCF's military toxics project and its environmental justice project were spun off as independent projects, Cohen said. "Dissolving NTCF and letting these organizations go their own way allowed the movement to be better served and concentrate resources more strategically." The Jobs and Environment Campaign, a product of the split now called the Environmental and Economic Justice Project (EEJP) is located near Boston and has become a training center for environmental justice activists. Quotes from Cathy Hinds, et al., *The National Toxics Campaign: Some Reflections, Thoughts for the Movement* (1994).

7. Quoted in Gottlieb, *Forcing the Spring,* p. 119.

8. *Everybody's Backyard* (Newsletter of CCHW), Spring 1993.

9. CCHW Annual Report, 1993.

10. Quoted in William Greider, *Who Will Tell The People: The Betrayal of American Democracy* (New York: Simon and Schuster, 1992), p. 168.

11. The Southern California Waste Management Forum described "The NIMBY syndrome" as "a public health problem of the first order . . . a recurring mental illness which continues to infect the public. . . . Organizations which intensify this illness are like the viruses and bacteria which have, over the centuries, caused epidemics such as the plague."

12. Interview with Gibbs.

13. Ibid.

14. Peter Montague, *Rachel's Hazardous Waste News,* vol. 104 (1989).

15. One corporate strategy to combat NIMBYism is acronymed GUMBY (Gotta Use Many Backyards). By proposing several different locations for a potentially hazardous facility, the company hopes to get the communities fighting among themselves over the final site.

16. Interview with Crockett, September 11, 1994.

17. Quoted in Margaret Kriz, "Shades of Green," *National Journal,* July 28, 1990, p. 1829.

18. Interview with Cindy Zipf.

19. Rose Gutfeld, *Wall Street Journal,* August 20, 1992, p. B3.

20. Karl Grossman, "Environmental Justice," *E* Magazine, May–June 1992, pp. 30–35.

21. General Accounting Office, *Siting Hazardous Waste Landfills and Their Correlation with the Racial and Economic Status of Surrounding Communities* (Washington, D.C.: Government Printing Office, 1983); Richard Moore, "Toxics Race and Class: The Poisoning of Communities," speech delivered March 18, 1991, Washington, D.C.

22. An attempt to discredit the UCC and GAO findings appeared in 1994. Researchers at the University of Massachusetts examined the siting of hazardous-waste treatment and storage facilities. The study, funded by the nation's largest waste handler, Chemical Waste Management, Inc. (Chem Waste), concluded, not surprisingly, that the UCC and GAO figures were incorrect. The newer study, however, contained an important flaw. It examined only Standard Metropolitan Statistical Areas (SMSAs, areas that contain a city of at least 50,000), whereas at least 15 percent of all hazardous-waste facilities are located in rural areas outside of SMSAs. The study left out some of the largest facilties, most notably those in Emmelle, Alabama, Pinewood, South Carolina (both predominantly African American communities), and Kettleman City, California (a largely Latino community). The disposal sites in Emmelle and Kettleman City, the two largest in the United States, are both owned by Chem Waste.

23. Marianne Lavelle and Marcia Coyle, "Unequal Protection." This award-winning investigative report appeared in the September 21, 1992, *National Law Journal,* p. S2.

24. Ibid.

25. D. R. Wernette and L. A. Nieves, "Breathing Polluted Air: Minorities Are Disproportionately Exposed," *EPA Journal* (March/April 1992); Lavelle and Coyle, "Unequal Protection," *National Law Journal*, p. S6.

26. Adam Walinsky, "The Blue Collar Movement," in *Earth Day—The Beginning: A Guide to Survival* (New York: Arno Press, 1970), p. 148–51.

27. "Playing With Fire" *Greenpeace* Magazine, editorial, 1990.

28. *Beyond the Green: Redefining and Diversifying the Environmental Movement* (Boston: The Environmental Careers Organization), 1992, p. 3.

29. Taylor, *Proceedings of the Michigan Conference on Race.*

30. Quoted in "Unequal Protection," p. S10.

31. Bullard, "Beyond Ankle Biting: Fighting Environmental Discrimination Locally, Nationally and Globally," *The Workbook,* vol. 16, no. 3 (Fall 1991).

32. David Helvarg, *The War Against the Greens* (San Francisco: Sierra Books, 1994), p. 47.

33. Dirk Johnson, "New Foe of the Indians," *The New York Times,* January 11, 1992.

34. Before the initiative, EPA ranked 46th of the 50 largest federal government agencies in the hiring and promotion of racial minorities.

35. Greenpeace also responded to Concerned Citizens of South Central Los Angeles (CCSC) when the City of Los Angeles proposed building an incinerator in South Central. CCSC's pleas for help to other national organizations went unheard by all but the

NRDC. Greenpeace also assisted actions against toxic-waste proposals on the Rosebud Indian Reservation in South Dakota and in Alsten, Louisiana.

36. Interview with Dana Alston, February 1, 1994; Gibbs quoted in Giovanna di Chiro, "Defining Environmental Justice," *Socialist Review* (April 1992), p. 103.

37. Dana Alston and Anne Bastion, "New Developments in the Environmental Justice Movement," September 1993.

38. Richard Moore, "Toxics, Race and Class: The Poisoning of Communities," March 1991.

39. The memo quoted here was leaked to this author and other journalists by Congressman Henry A. Waxman (D-CA).

40. Philip Landrigan, Chair of the Department of Community Medicine at Mount Sinai School of Medicine, New York, provided this statistic.

41. Quoted in Eric Mann, "Environmentalism in the Corporate Climate," *Tikkun* Magazine (March 1990), p. 62.

42. Mann, "Environmentalism in the Corporate Climate."

43. Quoted in Karen Elizabeth Woods, "Condoms and Corporations: Perspectives and Action on Population Among Environmental Organizations" (M.A. Thesis, University of Montana: Missoula, 1994), p. 39.

44. Ibid.

45. Quoted in *The Village Voice,* October 4, 1994.

46. *Sierra Club Bulletin* (January 1973).

47. Interview with Fischer, September 9, 1994.

48. Woods, "Condoms and Corporations."

49. Garrett J. Hardin, *The Limits of Altruism: An Ecologist's View of Survival* (Bloomington: Indiana University Press, 1977).

50. Quoted by Marc Cooper, *The Village Voice,* June 23, 1992.

51. Ibid.

52. Ibid.

53. McCown quoted in *Sierra* Magazine, October 1994; Bullard, ed., *Confronting Environmental Racism: Voices from the Grassroots* (Boston: South End Press, 1993).

54. Greider, *Who Will Tell The People,* p. 215.

55. In Donald Snow, ed., *Voices from the Environmental Movement: Perspectives for a New Era* (Washington, D.C.: Island Press, 1992), p. x.

56. Quoted in Greider, *Who Will Tell the People,* p. 214.

57. Greider, *Who Will Tell the People,* p. 219.

58. Interview with Alston.

Chapter 7

1. Frank Clifford, *Los Angeles Times,* September 21, 1994, p. 22.

2. The League of Conservation Voters ranked Clinton seventh among the seven Democratic candidates that ran in the New Hampshire primary.

3. Babbitt said he was "leading a liberation force" when he took over the Department of the Interior from the Republicans.

4. *Newsweek,* November 1, 1993.

5. Quoted by Rick Russell, *The Nation,* March 27, 1989, p. 405.

6. In March 1993 Jay Hair of the National Wildlife Federation did make national news with his comment about the administration's relationship with the environmental movement: "What started out like a love affair turned out to be date rape." However, he quickly redeemed himself at the White House by actively lobbying his Green Group colleagues to support NAFTA, accusing those who did not go along with the administration and the agreement of raising "phoney environmental issues."

7. Personal interview with Adams.

8. Quoted in Ann Reilly Dowd, *Fortune,* September 19, 1994, p. 104.

9. Gore's book, *Earth in the Balance* (Boston: Houghton Mifflin, 1992), shows the author to have a sophisticated understanding of environmental issues ranging from global warming to environmental justice. In it he challenges the gospels of GNP and GDP and "blind devotion to laissez-faire reponses." He states that the World Bank should have its "feet held to the fire" and calls for a global Marshall Plan for environmental protection and restoration. To grassroot enviros, his role in administration decisions reeks of apostasy.

10. *Newsweek,* November 1, 1993, p. 32.

11. Babbitt was so proud of the deal he hiked deep into Everglades National Park with Florida Governor Lawton Chiles to sign the "Everglades Forever Act." Originally the bill was to be called "The Marjory Stoneman Douglas Act," after the 104-year-old woman to whom President Clinton had recently awarded the Medal of Freedom in honor of her lifelong struggle to protect the Everglades. Douglas was so disappointed with the bill she demanded that Chiles remove her name. Representatives of Defenders of Wildlife and local environmentalists who attempted to attend the signing ceremony were ordered to stay in their cars in a parking lot three miles from the site.

12. "Fast-track" meant that the U.S. Senate could only vote up or down to ratify the treaty. Members would have no power over the terms of the agreement.

13. It was, in fact, the CEOs of those organizations not the staff or members, who endorsed the renegotiated NAFTA. These leaders were Jay Hair, NWF; John Adams, NRDC; Fred Krupp, EDF; Peter Berle, National Audubon; Dr. Russel Mittermeier, Conservation International; and Katherine Fuller, World Wildlife Fund.

In addition, several corporations formed an ad hoc organization, USA*NAFTA, in support of the agreement. Eastman Kodak, a founder of the group, contributed $250,000 to the World Wildlife Fund in 1992 (*see* the WWF 1992 Annual Report). Bank of America, another strong backer, made a generous donation to NRDC the same year.

14. *The New York Times,* September 16 1993.

15. Copies of Hair's letter to McCloskey were sent to the president, the vice president, the secretary of the interior, Ambassador Mickey Kantor, EPA Adminstrator Carol Browner, the Green Group, the NWF Board of Directors, endowment trustees, and NWF executive and regional staffs. Copies were sent to this author by sources unknown.

16. Interview with John Adams. In my interviews with NRDC lawyers and scientists in San Francisco and Los Angeles, I found no one who supported NAFTA.

17. Ibid.

18. Copy of letter sent by Evans to Alex Cockburn of Petrolia, California. Hermack's response is from a personal interview with the author.

19. Interview with St. Clair.

20. Jeffrey St. Clair, "Meditations on a Done Deal," *Wild Forest Review* (November 1993), p. 14.

21. Quoted in *Audubon* magazine (September/October 1994), p. 103.

22. Robert Mitchell, Angela Mertig, and Riley Dunlap, "National Environmental Organizatons," *Society and Natural Resources,* vol. 4, p. 228.

23. Center for Responsive Politics. During the same period, all business PACs contributed $295.4 million.

24. Interview with Dan Becker.

25. James B. Dougherty, "The Advocacy Challenge," *Defenders* magazine (May/June 1991), p. 7.

26. James Monteith, "A Strategy of Capitulation," *Wild Forest Review* (November 1993), p. 17.

27. Interview with Hayes, August 18, 1994.

28. Survey of American Voters: Attitudes Toward the Environment, Environment Opinion Study, Inc., June 1991.

29. The invitation list included John Adams from NRDC; Jay Hair of the National Wildlife Federation; Fred Krupp of EDF; Jim Maddy of the League of Conservation Voters; Peter

Berle, President of National Audubon; Barbara Dudley, Executive Director of Greenpeace; Jane Perkins, CEO at Friends of the Earth; Carl Pope, Director of the Sierra Club, and Margaret Morgan-Hubbard from Environmental Action.

30. *PR Watch,* second quarter 1994.

31. Interview with Nelson, August 26, 1994.

32. Interview with Hayes, August 18, 1994.

Chapter 8

1. Dunlap and Mertig, "The Evolution of the United States Environmental Movement from 1970 to 1990," *Society and Natural Resources,* vol. 4, p. 211–12.

2. Francisco R. Sagasti and Michael Colby, *Eco-Development and Perspectives on Global Change and Developing Countries*

3. With Foreman was Bart Koehler, who had worked in the Wilderness Society's Wyoming office; Howie Wolke, from the Wyoming office of Friends of the Earth; Ron Kezar; and a six-foot-six former Yippie named Mike Roselle.

4. Abbey, *The Monkey Wrench Gang* (New York: Avon, 1976).

5. Bill Devall interview with David Foreman published in *Simply Living* (Australia).

6. Earth First! created something of an identity crisis for some traditional enviros who agreed with much of the movement's rhetoric but found its style and tactics puerile and embarrassing. Despite the danger of being identified with eco-saboteurs and guerrilla antics a few old timers found it in their hearts to cast a little holy water on the buckaroos and their theatrics. "They make us look reasonable." said Bob Hattoy, then of the Sierra Club, now of the Interior Department. David Brower, Michael Fischer, and even Bill Turnage, who was running the Wilderness Society when Foreman left it, have had kind words for Earth First!.

Closet support for the organization appears to have reached all the way into the heart of the environmental establishment. One prominent member of the EDF Board, who cautioned me he would lose his board membership and his job if I revealed his identity, told me he surreptitiously supports Earth First! "with larger donations than I make to EDF." Another EDF board member who supported Earth First! said, "I can't let anyone know I am affiliated with the group, so I give lifetime subscriptions of *Wild Earth* (Earth First!'s journal) to my family and friends."

7. Quoted in Robert Braile, "What the Hell Are We Fighting For," *Garbage* Magazine (Fall 1994), p. 35.

8. According to estimates of conservation biologists Reed Noss, Michael Scott, and Edward La Roe, the loss of virgin forest could be as high as 98 percent in the lower 48 states, including 99 percent of the longleaf pine in Florida and 99.95 percent of the beech-maple forest in Michigan.

9. *Wild Forest Review,* vol. 1, no. 1 (November 1993), p. 15.

10. Interview with Mahler.

11. Stephanie Mills, *Whatever Happened to Ecology?* (San Francisco: Sierra Club Books, 1989).

12. Interiew with David Orr.

13. Ibid.

14. Quoted in Frank Clifford, *Los Angeles Times,* September 21, 1994, p. A24.

15. Interview with Michael Fischer.

16. David Kalish, AP wire story. September 12, 1991.

17. Op Ed story, Pacific News Service, June 16, 1989.

18. Interview with Foreman, September 10, 1994.

19. Quoted in *Voices from the Environmental Movement: Perspectives for a New Era,* Don Snow, ed. (Washington D.C.: Island Press, 1992), p. 68.

20. Howard Hawkins, "The Politics of Ecology: Spinning the Web of Social Theory," *The Guardian* (Manchester, England), April 25, 1990.

21. Speech by Rachel Carson, 1964.

22. Commoner, *Making Peace with the Planet* (New York: Pantheon, 1990), p. 16.

23. Arne Naess, "The Shallow and Deep, Long Range Ecology Movement: A Summary," *Inquiry,* vol. 16, no. 1 (Spring 1973), p. 95.

24. Quoted in Bill Devall, "Deep Ecology and Radical Environmentalism," *Society and Natural Resources,* vol. 4, p. 251.

25. Ibid., p. 250.

26. Mills, *Whatever Happened to Ecology?,* p. 127.

27. Murray Bookchin, "Ecology and Revolutionary Thought," in *Post-Scarcity Anarchism* (Berkeley: Ramparts Press, 1971).

28. Quoted in Mills, *Whatever Happened to Ecology?* The seminal document of bioregionalism is probably Raymond F. Dassman's 1976 essay, "Bioregional Provinces," *Coevolution Quarterly* (Fall 1976), pp. 32–37. The idea was popularized by *Coevolution Quarterly* during the 1970s and is still promoted by the Planet Drum Foundation in San Francisco.

29. Seager, *Earth Follies* (New York: Routledge, 1993).

30. Warren, "Toward an Ecofeminist Ethic," *Studies in the Humanities* (December 1988), pp. 140–56.

31. Quoted in Harold Gilliam, "Eco Therapy," *This World* (*San Francisco Chronicle*), September 12, 1993, p. 14.

32. Interview with Vander Ayn.

33. Roush quoted in Snow, *Voices from the Environmental Movement,* p. 15.

34. Stephen Viederman, "Sustainable Development: What Is It and How Do We Get There?" *Current History,* vol. 92 (April 1993), p. 180.

35. Interview with Gibbs.

36. Newman and Schneiderman, "Market Leftism," *Crossroads,* May 1994.

37. John O'Connor, "The Promise of Environmental Democracy," *Toxic Struggles,* 1993, p. 54.

38. Josh Reichert's analysis is quoted from various interviews with him in September 1994.

39. Quoted in Robert Braile, "What the Hell Are We Fighting For?" *Garbage* Magazine (Fall 1994), p. 32.

40. Bullard's books are *Dumping in Dixie: Race Class and Environmental Quality* (Boulder: Westview Press, 1990), *Confronting Environmental Racism: Voices from the Grassroots* (Boston: South End Press, 1993), and *Invisible Houston: The Black Experience in Boom and Bust* (College Station: Texas A&M University Press, 1987).

41. Quoted in Olaus J. Murie, "Wilderness Is for Those Who Appreciate," *The Living Wilderness,* vol. 5, no. 5 (July 1940), p. 5.

42. Robert Gottlieb, *Forcing the Spring: The Transformation of the American Environmental Movement* (Washington, D.C.: Island Press, 1993), p. 19.

43. Michael Frome, "Heal the Earth, Heal the Soul," in *Crossroads: Environmental Priorities for the Future,* Peter Borrelli, ed. p. 249.

Epilogue

1. According to World Watch, 70 percent of the world's 9,600 bird species are declining, and 1,000 of the species are threatened with extinction.

2. *Time* Magazine, February 3, 1988, p. 31.

3. "In the time it takes Joe Montana to throw a touchdown pass, 100 football fields of rainforest will be clearcut," said David Brower.

4. Quoted in Kirkpatrick Sale, *The Green Revolution: The American Environmental Movement, 1962–1992* (New York: Hill and Wang, 1993), p. 61.

Appendix A

Principles of Environmental Justice

Adopted at the First National People of Color Environmental Leadership Summit, October 24–27, 1991, Washington, D.C.

Preamble

We, the people of color, gathered together at this multinational People of Color Environmental Leadership Summit to begin to build a national and international movement of all peoples of color to fight the destruction and taking of our lands and communities, do hereby re-establish our spiritual interdependence to the sacredness of our Mother Earth; to respect and celebrate each of our cultures, languages and beliefs about the natural world and our roles in healing ourselves; to insure environmental justice; to promote economic alternatives which would contribute to the development of environmentally safe livelihoods; and, to secure our political, economic and cultural liberation that has been denied for over 500 years of colonization and oppression, resulting in the poisoning of our communities and land and the genocide of our peoples, do affirm and adopt these Principles of Environmental Justice:

1. *Environmental justice* affirms the sacredness of Mother Earth, ecological unity and the interdependence of all species, and the right to be free from ecological destruction.

2. *Environmental justice* demands that public policy be based on mutual respect and justice for all peoples, free from any form of discrimination or bias.

3. *Environmental justice* mandates the right to ethical, balanced and responsible uses of land and renewable resources in the interest of a sustainable planet for humans and other living things.

4. *Environmental justice* calls for universal protection from nuclear testing and the extraction, production and disposal of toxic/hazardous wastes and poisons that threaten the fundamental right to clean air, land, water, and food.

5. *Environmental justice* affirms the fundamental right to political, economic, cultural and environmental self-determination of all peoples.

6. *Environmental justice* demands the cessation of the production of all toxins, hazardous wastes, and radioactive materials, and that all past and current producers be held strictly accountable to the people for detoxification and the containment at the point of production.

7. *Environmental justice* demands the right to participate as equal partners at every level of decision-making including needs assessment, planning, implementation, enforcement and evaluation.

8. *Environmental justice* affirms the right of all workers to a safe and healthy work environment, without being forced to choose between an unsafe livelihood and unemployment. It also affirms the right of those who work at home to be free from environmental hazards.

9. *Environmental justice* protects the right of victims of environmental injustice to receive full compensation and reparations for damages as well as quality health care.

10. *Environmental justice* considers governmental acts of environmental injustice a violation of international law, the Universal Declaration On Human Rights, and the United Nations Convention on Genocide.

11. *Environmental justice* must recognize a special legal and natural relationship of Native Peoples to the U.S. government through treaties, agreements, compacts, and covenants affirming sovereignty and self-determination.

12. *Environmental justice* affirms the need for urban and rural ecological policies to clean up and rebuild our cities and rural areas in balance with nature, honoring the cultural integrity of all our communities, and providing fair access for all to the full range of resources.

13. *Environmental justice* calls for the strict enforcement of principles of informed consent, and a halt to the testing of experimental reproductive and medical procedures and vaccinations on people of color.

14. *Environmental justice* opposes the destructive operations of multi-national corporations.

15. *Environmental justice* opposes military occupation, repression and exploitation of lands, peoples and cultures, and other life forms.

16. *Environmental justice* calls for the education of present and future generations which emphasizes social and environmental issues, based on our experience and an appreciation of our diverse cultural perspectives.

17. *Environmental justice* requires that we, as individuals, make personal and consumer choices to consume as little of Mother Earth's resources and to produce as little waste as possible; and make the conscious decision to challenge and reprioritize our lifestyles to insure the health of the natural world for present and future generations.

Appendix B

Sierra Club Centennial Address

by Michael L. Fischer, Executive Director
May 28, 1992, Harpers Ferry, West Virginia

I. Introduction

Today, we celebrate an anniversary during a time of unprecedented environmental threat. The urgency of our mission has never been greater than during this decade of the 1990's. This is a time for celebration, yes! But also a time for anger, for worry, for hope.

We are marking ten decades of activism and sacrifice, one hundred years of struggle, a centennial of service to the highest end we recognize. And, we're looking forward to beginning our second century by building a new (long-overdue) partnership, which I will describe.

Before we get to our celebration, I want to talk about something that happened at the very creation of the United States that places our fight in context. I want to talk about two tragic omissions made by Thomas Jefferson when he wrote the Declaration of Independence: one that has made the environmental struggle necessary; and another that will probably shape its future for at least the next one hundred years.

The first omission is in the category of what Jefferson called "unalienable rights." Mr. Jefferson lived in a world that seemed divinely constructed to support and sustain human beings. It was impossible for him to imagine—as it was impossible for generations as recent as that of our parents to imagine—a world in which the quotidian human routines of war, settlement, business, and daily life could threaten the living balance of an entire planet.

So in his magnificent catalog of unalienable rights we find, to our abiding sorrow, no mention whatsoever of the right to breathe clean air; or the right to drink pure water; or the right to live unassailed by toxins, unbombarded by nuclear radiation, unsuffocated by poisonous atmosphere, unexposed to the ferocious and unfiltered energies of Space, unthreatened by global changes in climate. He included none of these rights, and we have lived long enough to regret it.

His second omission is in the category of "self-evident truths," and here Jefferson didn't miss by much. He wrote that "all Men are created equal," very much in the Enlightenment's mainstream. But he neglected to specify exactly who these "Men" (and women, Tom?) were, thus leaving the field open for those who would construe, assert, and ultimately be willing to die for the belief that the men gathered here today, metaphorically lit by the one-month-old flames of South Central Los Angeles—and seeing in our mind's eye the burning buildings and violence of Watts in 1965, Newark in 1968, Miami in 1980, and all the other times people of color rose up in despair against a system which crushes them—we can see with unusual clarity the price we have paid for his lack of specificity.

But what do these omissions have to do with the centennial of the Sierra Club?

The answer is "Everything." They define its past and its future.

II. Environmentalism and the First Omission

The first omission can be regarded as the headwaters of the environmental movement. *It made environmentalism necessary.* Had the right to a clean biosphere been mentioned in the Declaration, it might have been given the same Constitutional protections now accorded the right to bear arms, the right not to quarter troops in your home, and other such niceties. Had it been included in the Declaration, John Muir and his friends would never have had to sit down in a San Francisco office on May 28th, 1892—one hundred years ago today—to draft and sign the articles founding the Sierra Club.

But environmental rights were *not* included. And John Muir, and his friends, and all of his spiritual descendants have been forced to act to remedy that omission. Their first fight was defeating a proposed reduction of Yosemite National Park.

Since that effort, the club has fought a One Hundred Years War for responsible stewardship of our planet, a fight whose consequences will forever outweigh those of the one-hundred-year squabble between England and France in the fourteenth century. We have won some brilliant victories. We have suffered painful defeats. But we have never given in to the forces of those who worship growth, and who seem to regard the Earth as little more than a platform for the display of ever-larger and more destructive forms of commercial solipsism:

- In 1905, we fought to get Yosemite returned to federal management, and won.

- In 1911, we fought the battle to establish the Devil's Postpile National Monument, and won.

- In 1923, we were instrumental in preventing the Kings River in the Sierra Nevada from being dammed.

- In 1935, we fought to establish King's Canyon National Park, an effort successfully concluded five years later.

- In 1948, we helped defeat a dam that would have flooded 20,000 acres of Glacier National Park.

- In 1951, we introduced the concept of federal legislative protection for wilderness areas, a powerful weapon that we and many other conservation groups have used to tremendous effect.

- In 1963, we initiated the fight to keep federal agencies from damming and flooding parts of the Grand Canyon.

- From the Yazoo River in Mississippi, to the Oachita National Forest in Arkansas, to the Everglades, to the Adirondacks, the Sierra Club has been there and is there now.

- In 1973, we launched the campaign that successfully defended the Clean Air Act from the ravages of the auto industry.

• In 1980, we won the fight to establish the "Superfund" law designed to help clean up toxic waste.

• And in 1991, in concert with the Gwichin people, we led the effort to defeat the Johnston-Wallop Energy Bill, the effort of our self-proclaimed "environmental president" and his Bronze-Age advisers to gut the Clean Air Act, resuscitate nuclear power, destroy the Arctic National Wildlife Refuge, and sacrifice large portions of our riparian lands. All to serve the short-term interests of those who today manipulate our political and economic systems, to the detriment of our children—our future.

Every decade of our first hundred years has seen conservation victories of national significance. And it is time to celebrate! But the price of environmental quality is eternal vigilance, so the past is but prologue.

III. Civil Rights and the Second Omission

Just as the first of Mr. Jefferson's omissions was the headwaters of the environmental movement, the second of his tragic omissions can be regarded as the headwaters of the civil rights movement.

It is appropriate that we discuss racism and civil rights here today for two reasons. First, because it was within a few miles of this place, 133 years ago, that Old Osawatomie Brown struck his famous blow for *his* self-evident truth.

When John Brown and his 22-man party attacked the federal arsenal at Harpers Ferry, his immediate goal was to equip an army of liberation. But this raid, put down by Colonel Robert E. Lee, precipitated an avalanche of blood and iron, a civil war, fought ultimately to rectify Mr. Jefferson's second omission—to include people who are not white in the Declaration's definition of humankind.

Ever since the ratification of the 13th amendment in 1865, the civil rights movement has fought and won its battles by itself. The Civil Rights Act was passed by a shaken Congress in 1866. The NAACP won its great triumph in Brown vs. Board of Education in 1954. That was followed by passage of the Voting Rights Act of 1964, and the establishment of the Equal Employment Opportunity Commission soon thereafter. All of this happened with very little assistance from nongovernmental organizations concerned with other social issues.

But by many measures, the civil rights envisioned by Jefferson remain denied to many Americans today. This certainly holds true if you think in terms of the environmental rights which we at the Sierra Club have been fighting to protect these last 100 years.

So let's review where we are. We have seen that America has suffered from two omissions from the Declaration of Independence, and that the two omissions have given rise to two major grass-roots movements in the 19th and 20th centuries. But until now, these two movements have been almost entirely separate—like two neighboring rivers draining separate watersheds.

And this brings us to the second reason racism and civil rights concern us today. The environmental movement and the civil rights movement have so far been effective acting alone. But in the future, it will be impossible for either movement to get as much done separately as they could if we joined forces.

Let the word go forth from this time, from this place: My purpose in being here on this Centennial Day is to invite a friendly takeover of the Sierra Club by people of color.

IV. Identity of Interest

Why should the Sierra Club want such a takeover? And why should people of color be interested in executing it?

The answers to these two questions are written on the landscapes of poverty in every region of this country and around the world. The fact is that while the rich are busy creating the poisonous byproducts of our economic system, the poor have to live submerged in them. The world's high-risk toxic environments are not in Georgetown, or Greenwich, or Malibu. They are in the inner cities, or in the towns where poor, disenfranchised people of color live. On the South Side of Chicago. In Martinez, California. In Harlem. On the Pine Ridge Indian Reservation in South Dakota. In the Liberty City section of Miami. In East St. Louis. In West Dallas. In Cancer Alley, between Baton Rouge and New Orleans.

- Living near major highway corridors in inner cities has been found to elevate the levels of lead in the bloodstream. Who lives near those corridors?

- Working in fields where pesticide use is high significantly increases the risk of systemic poisoning for farm workers. Who works in those fields?

- Living in old-line manufacturing zones means living in an atmosphere polluted by toxic waste incinerators, coal-burning power plants, chemical plants, and other facilities too dirty or dangerous to be placed in the suburbs. Who lives in those zones?

People of color do. People of color bear a disproportionately large share of the economy's toxic burden. Sixty percent of African American and Latino populations in the U.S. live in communities with one or more toxic waste sites. Sixty percent of the country's largest commercial hazardous waste landfills are situated in predominantly black or Latino neighborhoods.

Yet these same people reap a tiny share of whatever benefits the economy might offer.

This is where the Sierra Club and the civil rights movement share an absolute identity of interest. We have dedicated ourselves to fighting the degradation of the environment that will kill us all if it proceeds at the rate advocated by the "What, Me Worry?" Republican Right. The civil rights movement has dedicated itself to fighting the systemic injustice suffered by people of color every day of their lives.

The point is this: If people of color bear the brunt of our economy's poisons, what is that but just another form of injustice, of racism? In fact, we see in the lives of the nation's poor the final confluence of racial injustice and environmental degradation to create environmental injustice. And I believe the struggle for environmental justice in this country and around the globe must be the primary goal of the Sierra Club during its second century.

This is the battleground where the environmental movement and the civil rights movement can join forces. *Must* join forces, really. Both groups are being compelled by events to act. People of color must adopt the environmental agenda to survive. The Sierra Club must embrace multiculturalism. We have nothing at stake but our moral and ethical integrity.

The Club exists to influence public policy. This is where our enlightened self-interest comes in. In the United States, the decision-making bodies (city councils, the courts, state legislatures, Congress) are, thank God, becoming more ethnically diverse. In the rest of the world, where all of us have a stake in the development options they choose, they are, by definition, culturally and racially diverse.

So we are faced with a choice: Will we remain a middle-class group of backpackers, overwhelmingly white in membership, program, and agenda—and thus condemn ourselves to losing influence in an increasingly multicultural country? Or will we be of service to, of relevance to people of color, combine forces, and strengthen our efforts at our chapter and group level, especially in the localities where the environmental and civil rights battles are going to be lost or won?

We must build that bridge now. We must refuse to become irrelevant.

I should note that, as we are attempting to build the bridge, others with apparently cynical and sinister purpose are attempting to tear it down. Just as campaign operatives identified Willie Horton as a target for dividing America, just as operatives identified the Murphy Brown character as opportunity for a cheap shot at women, an EPA political appointee in February identified a draft EPA report on environmental equity as an opportunity to divide traditional environmental organizations from the emerging movement for environmental justice, to split off the constituencies into rival camps. We all quickly saw through this ploy and called a halt to this destructive venture.

Two weeks ago the President himself took a whack at the bridge between environmentalists and our polluted communities when he ordered the Environmental Protection Agency to publish an illegal regulation for states' programs to issue air pollution permits. The regulation precludes the full public participation envisioned by Congress in the all-important process of getting a permit, and it allows industry to essentially permit themselves under the guise of "minor permit adjustments."

Now where are these self-permitted polluting facilities disproportionately located? In what neighborhoods will the permits issued with virtually no public comment be clustered? Communities of color, of course: already overly polluted, already disenfranchised and cut out of the debate in so many ways. Insult upon insult.

Again, the Sierra Club joins communities of color in exposing and fighting this cynical manipulation of public policy. Gutting clean air permits that won't go into effect for several years will do nothing for the pre-election economy. In fact, planning for tight permits would fuel the economies of industrial communities.

Tight permits can be used to produce permanent, good jobs in industrial communities in open debate and consensus-building about how to sustain productive, but clean industrial economies. This process will attract capital to the facilities and nearby service providers. It will also curtail human suffering, health damage and health care costs. I know this from my discussions with union leaders and community activists in Lousiana's Cancer Alley. I say that we must join with communities of color to use environmental protection as a vital step in industrial and economic revitalization.

I recently attended one of the plenary sessions of the first National People of Color Environmental Leadership Summit, organized by Reverend Ben Chavis—and that really opened my eyes. I was in a room full of 600 environmentalists, all of them informed, hardworking, courageous, and committed, and fewer than 100 of us white. I was elated, inspired, and moved by the experience. But how are we going to work together effectively if in the environmental movement we do not broaden our own membership? It's good news to learn that 7% of our members tell us that they are people of color. It's better news to learn that 12% of our members under 30 years old are people of color. But, there is more, much more, that we must do.

Last year I spoke to Qu Geping, China's EPA administrator, about his concerns about the future. He spoke of the fact that his country's energy supply is soft coal, and that the program of modernization will have refrigerators in most Chinese households by the middle of the next century. He clearly sees the potential impact on China's future of these two facts, saying simply: "I fear for my children. Regardless of what the developed nations do to limit greenhouse gases, I fear for my children."

We know some of the steps Mr. Qu should take if he wants to avert this catastrophe. But China is a culture radically different from our own. How are we going to have any influence on them if we have not demonstrated a commitment to multiculturalism within our own organization?

The need to retain our credibility and expand our influence in a diverse world is behind the Sierra Club's Ethnic and Cultural Diversity Initiative, an effort to:

- Increase the ethnic and cultural diversity among our leaders, members, and volunteers;
- Increase the ethnic and cultural diversity of our staff and independent contractors; and
- Increase Sierra Club interaction with ethnically diverse groups on environmental issues.

As part of that effort, we have also used these criteria to decide whether to fund 50 grant requests from Sierra Club chapters and groups for environmental outreach programs. Among the programs funded were:

- Hiring a person of color as a grassroots organizer in the South, which has become a wasteland where the well-being of the poor and people of color is subordinated to the demands of business and the laxity of local governments;
- Working with the Coalition Against Childhood Lead Poisoning in Baltimore, where one-half of the children of color screened for lead poisoning had elevated lead levels. This grant will be used to create educational materials, develop a media campaign, and hire an organizer to keep the community aware of the dangers of lead;
- Convening an agenda-setting meeting of our Great Lakes volunteer and staff leaders with Native American environmental leaders from across the ecosystem. Indian people who hunt and fish for much of their food are particularly at risk to toxics. In cooperation with Indian leaders, we'll survey the tribes to identify specific toxics issues and then develop a realistic, cooperative plan to tackle them, learning across cultures as we do.

The grants and the diversity initiative will begin the effort. But the battle for environmental justice will be fought and won at the local level. And the Sierra Club is absolutely the best-positioned organization to influence those decisions.

We have in existence now 425 local chapters and groups in almost every corner of this country. *Every one of them* can begin to establish coalitions and networks with local advocacy groups for people of color when decisions affecting local environments come before city councils and county boards of supervisors. *Every one of them* has an open membership policy, and *every one of them* welcomes new volunteer activists. This is where the friendly takeover I call for can best take place. The power, the commitment, the courage, the effectiveness, the history of victories which Sierra Club chapters and groups have proudly built over the decades is ready for expanded service.

In other words, the infrastructure is already in place. We (in both movements) just have to recognize the need to use it.

V. Conclusion

From Thomas Jefferson's two omissions 216 years ago, from those two seemingly insignificant sources, two major grass-roots movements have sprung like two long and powerful, separate rivers. Our histories have been impressive; but the future needs us like never before.

And now those two rivers are flowing together, like the Shenandoah and the Potomac Rivers, just over that ridge here at Harper's Ferry, are joining forces like the streams of Yosemite described so beautifully by John Muir:

> It seems strange that visitors to Yosemite should be so little influenced by its novel grandeur, as if their eyes were bandaged and their ears stopped. Most of those I saw yesterday were looking down as if wholly unconscious of anything going on about them, while the sublime rocks were trembling with the tones of the mighty chanting congregations of waters gathered from all the mountains round about, making music that might draw angels out of heaven.

The music we make as we join forces may not draw angels out of heaven. But it may draw people out of their complacency, and cause them to see what we have done—cynically or unintentionally—to communities of color across our land. It may cause people to rise up and demand justice—environmental justice—for all.

I hope it does. The vibrant first century of the Sierra Club is now ended, and the second century beckons us—all of us. There is no more time to wait.

Appendix C

The Wise Use Agenda

A Task Force Report Sponsored by the Wise Use Movement
by Alan M. Gottlieb

Section I: The Top Twenty-Five Goals

1. *Initiation of a Wise Use Public Education Project* by the U.S. Forest Service explaining the wise commodity use of the national forests and all federal resource lands. An important message is that the federal deficit can be reduced through prudent development of federal lands. A public outreach action plan and implementation shall be accomplished using print and electronic media to reach the broadest possible public with the commodity use story of the national Forest System.

2. *Immediate wise development of the petroleum resources of the Arctic National Wildlife Refuge (ANWR)* in Alaska as a model project showing careful development with full protection of environmental values.

3. *The Inholder Protection Act.* To provide congressional recognition to the lawful status as property owners of inholders within all federal areas. The United States shall irrevocably recuse itself from all eminent domain power over inholdings. The Act should repeal the General Condemnation Act of 1888 and the Declaration of Taking Act of 1933. Acquisition of private land from federal inholders shall henceforth take place only with the uncoerced agreement of the inholder.

4. *Passage of the Global Warming Prevention Act* to convert in a systematic manner all decaying and oxygen-using forest growth on the National Forests into young stands of oxygen-producing, carbon dioxide-absorbing trees to help ameliorate the rate of global warming and prevent the greenhouse effect. The federal government shall also help fund and coordinate urban tree planting on all federal property as part of this critical program.

5. *Creation of the Tongass National Forest Timber Harvest Area* in Alaska limiting timber harvest to only 20 percent of the total national forest's 17 million acres, or 3 million acres, for the next century, allowing only 30,000 acres to be harvested per year. The Tongass Timber Harvest Area is to be the first unit of the National Timber Harvest System designed to promote proper economic forestry practices on the federal lands as outlined in Goal Number Ten.

6. *Creation of a National Mining System* by Congressional authorization, to embody all provisions of the General Mining Act of 1872 with the added provision that all public lands

including wilderness and national parks shall be open to mineral and energy production under wise use technologies in the interest of domestic economies and in the interest of national security.

7. *Passage of the Beneficial Use Water Rights Act* to embody all the provisions of the Water Act of 1866 with the added provisions that Congress shall recognize as sovereign the rights of states in all matters related to the regulation and distribution of all waters originating in or passing through the states, and that the federal government shall not retain "reserved" or other federal property rights in waters arising on federal lands for which it cannot demonstrate beneficial use.

8. *Commemorate the 100th Anniversary of the founding of the Forest Reserves by William Steele Holman* who introduced the Section 24 rider to the Forest Reserve Act of 1891. This commemoration shall emphasize the Homestead Act of 1888 from which the Section 24 rider was derived and the commodity use and homestead settlement intent behind the law that created the national forests.

9. *The Rural Community Stability Act* shall give statutory authority to enable the U.S. Forest Service to offer a reasonable fraction of the timber on each Ranger District in timber sales for the sole and proper purpose of promoting rural timber-dependent community stability, exempt from administrative appeal.

The Forest Service shall offer an adequate amount of timber from each Ranger District in the United States National Forest System to meet market demands up to the biological capacity of the district and sell such timber only to local logging firms and milling firms.

The first timber sales allowed under the provisions of this act shall be those designed to recapture accrued undercuts from previous years when the annual allowable harvest level has not been achieved.

10. *Creation of a National Timber Harvest System* by Congressional authorization, to identify and preserve for commodity use those timberlands suitable for sustained yield timber growth. Repeals non-declining, even-flow strictures of the Forest and Range Renewable Resource Planning Act of 1974 as amended by the National Forest Management Act of 1976. Identifies wise use technologies acceptable to harvest timber in the interest of domestic economies and in the interest of national security. Applies the Multiple Use - Sustained Yield Act of 1960 provisions that lands will be managed in "not necessarily the combination of uses that will give the greatest dollar return of the greatest unit output," so that no timber harvest plan may be identified as "below cost." No enactment is to impair the agency's ability to manage the National Timber Harvest System for timber harvest.

The Tongass Timber Harvest Area should be the first dedicated single-use timber harvest area in America's National Timber Harvest System, to consist of the 3 million acres identified by the TLMP II planning process as suitable for growing commercial timber. Such areas should permit Multiple Use recreation where feasible. The Tongass timber industry should have all its former logging lands restored to logging status for the 100 year rotation so that at least 20 percent of the Tongass is scheduled for timber harvest over the next 100 years.

11. *National Parks Reform Act,* to create protective agencies for our natural heritage of a size conducive to responsible management and accessible to congressional oversight. Creates within the Department of the Interior, under authority of the Assistant Secretary for Fish and Wildlife and Parks four separate agencies each with its own director responsible for management of our current over-sized and jumbled national park system. Reorganizes the National Park Service, with new management responsibility limited to only those units officially designated "national parks" and "national monuments" in the "natural" category; creates the National Urban Park Service with management responsibility for all units of the park system in urban settings designed primarily for contemplation, enlightenment

or inspiration, such as the National Capitol Parks; creates the National Recreational Park Service with management responsibility for all National Recreation Areas of the park system and other units primarily used for recreational purposes; creates the National Historical Park Service with management responsibility for all national historic parks and similar units of primarily historic interest.

The present National Park Service with its domain in excess of 80 million acres has grown into a bureaucracy so huge and powerful that it can ignore the public will, the intent of Congress and direct orders of the Secretary of the Interior with impunity. Such concentrated power cannot be allowed to persist within a representative form of government. This Act will separate out from the present conglomeration of diverse units four different kinds of national heritage lands that have previously been lumped together into a single vast and unresponsive agency. The new arrangements will group together those that are naturally similar for appropriate management to protect the essential character of each different kind of park.

MISSION 2010: Adequate Park Visitor Accommodations. A major thrust should be made to properly accommodate the increased visitor load on our parks through a 20-year construction program of new concessions including overnight accommodations, classic rustic lodges, campgrounds and visitor service stores in all 48 national parks, with priority given to Great Smoky Mountain, Everglades, Rocky Mountain, Big Bend, Canyonlands, Sequoia, Redwoods, North Cascades, Denali, and Theodore Roosevelt. Concession restoration should begin immediately in Yellowstone (West Thumb). The lodge at Manzanita Lake in Lassen Volcanic National Park, which was demolished by the National Park Service, shall be rebuilt in replica on its original site and become the first project of Mission 2010, to become known as the Don Hummel Memorial Lodge honoring the late outstanding leader of the national park concession movement. The Concession Policy Act of 1965 should be extended to all facilities of the proposed four park services.

Appropriate overnight visitor facilities should be constructed in all national monuments, national recreation areas, and major historical areas. Policies that exclude people shall be outlawed. The possessory interest of the private concessioner firms now serving the visiting public should be maximized. Private firms with expertise in people-moving such as Walt Disney should be selected as new transportation concessioners to accommodate and enhance the national park experience for all visitors without degrading the environment.

All actions designed to exclude park visitors such as shutting down overnight accommodations and rationing entry should be stopped as inimical to the mandate of Congress for "public use and enjoyment" in the National Park Act of 1916.

12. *Pre-Patent Protection of Pest Control Chemicals.* The patent clock on newly discovered pest control chemicals should start running only after government-imposed regulation-compliance requirements have been met. Since the testing period for new chemical approvals typically exceeds three years, during which the owner of the chemical can realize no income on investment, it is only fair that patents run from the time an innovation becomes marketable, yet pre-patent protection should be granted as a matter of governmental duty.

13. *Create the National Rangeland Grazing System.* Congress should authorize a National Rangeland Grazing System on all federal lands presently under permit according to the terms of the Taylor Grazing Act [43 U.S.C. 315-315(o)], or managed as rangeland under the Federal Land Policy and Management Act of 1976 [43 U.S.C. 1701-1782] or other applicable rangeland statutes, and which (1) generally contains split estate values of privately owned possessory interests in the Federal lands, including but not limited to: water rights, range rights, privately owned range improvements such as roads, fences, stock watering facilities, ranch houses, cook houses, and bunk houses, (2) is rendered more valuable by the

contribution of commensurable private land, (3) is biologically suited to grazing by either intensive or extensive livestock management methods, and (4) may also be available to multiple use for purposes including but not limited to hunting, hiking, motorized recreation, watershed management, wildlife management, timber harvest and mineral management but no application shall impair the operation of the rangeland as livestock grazing areas.

14. *Compassionate Wilderness Policy.* The Veterans and Handicapped Wilderness Provision should be enacted by statute to allow motorized wheel chairs into all Wilderness Areas in the National Wilderness Preservation System.

15. *National Industrial Policy Act.* Enact the following provision: "all agencies of the Federal Government shall include in every recommendation or report on proposals for legislation and other major Federal actions significantly affecting the quality of the human environment, a detailed statement by the responsible official:

(vi) the economic impact of delaying or denying the proposed action,
(vii) the economic benefits of immediately going forward with the proposed action.

16. *Truth In Regulation Act.* In all agency plans that presently combine the production costs with overhead costs of a Federal action such as the offering of a timber sale on a national forest, all non-production costs shall be identified separately and conspicuously, including costs of writing the NEPA Environmental Impact Statement, costs of complying with environmental regulations on the ground, costs of government buildings, vehicles, and utilities required to complete the plan, and salaries and benefits of all agency staff employed in the project.

17. *Property Rights Protection.* Railroad easements when abandoned by the original or successor railroad operating company, shall revert to the underlying adjacent property owner. No easement shall be given by government decree to a "Rails-to-Trails" program without payment of just compensation plus money damages for loss of economic opportunity.

18. *Endangered Species Act Amendments.* The Endangered Species Act shall be amended to specifically classify the appropriate scientifically identified endangered species as relict species in decline before the appearance of man, including non-adaptive species such as the California Condor, and endemic species lacking the biological vigor to spread in range such as the wildflower Pipers harebells of the Olympic Mountains. Federal projects designed to protect species identified as relicts shall require a report stating all costs, separate and cumulative, of protecting the relict species, including computations of lost economic opportunities for projects denied because of the relict species.

All costs associated with mitigation and protection efforts required by federal law to protect endangered species shall be fully identified, separated from accounting statements and documented and made available for public inspection in an annual report to the Congress to be filed by the secretaries of affected departments.

Hiding, disguising or willfully concealing the existence of an endangered species protection cost shall be a felony malfeasance of office subject to severe penalties of fine and imprisonment.

19. *Obstructionism Liability.* Any group or individual that challenges by litigation an economic action or development on federal lands and subsequently loses in court shall be declared "not acting in the public interest" and shall be required to pay to the winner the increase in costs for completing the project plus money damages for loss of economic opportunity.

Congress should provide for obstructionists to indemnify American industry against harm when they use the law to delay economic progress. The law must require that those who bring administrative appeals or court actions against timber harvest plans, mining plans, grazing plans, petroleum exploration or development plans or other commodity uses of federal lands shall post bonds equivalent to the economic benefits to be derived from the challenged harvest plus cost overruns caused by delay. If the appellant or plaintiff loses, payment in full is to be made to the defendant in proportion to his losses and expenses.

20. *Private Rights in Federal Lands Act.* Congress should enact measures which recognize that private parties legitimately own possessory rights to timber contracts, mining claims, water rights, grazing permits and other claims that are recognized by the several states and by Internal Revenue Service estate tax collection policy as valuable private property rights. Establishes the principle of the Private Domain in Federal Land.

21. *Global Resources Wise Use Act.* Congress should enact a policy measure that explicitly recognizes the shrinking relative size of the total goods sector of our world's economy and takes steps to insure raw material supplies for global commodity industries on a permanent basis. Should include free trade measures and incentives for developing nations that favor private enterprise. Should provide for technology exchange of wise use methods and a global data bank of technical information on sustainable resource development processes including prevention and cleanup techniques.

22. *Perfect the Wilderness Act.* The National Wilderness Preservation System must be re-assessed and reclassified into more carefully targeted categories according to the actual appropriate use, including:

1) Human Exclosures, areas where people are prohibited altogether, including wildlife scientists who frequently harass to death the very animals they are supposed to protect;

2) Wild Solitude Lands, managed exactly as present Wilderness areas;

3) Backcountry Areas, which allow widely spaced hostels, primitive toilets to prevent unsanitary conditions that prevail today along Wilderness trails, and higher trail standards to prevent the trail erosion that plagues current Wilderness areas;

4) Frontcountry, to allow primitive and developed campsites, motorized trail travel and limited commercial development;

5) Commodity Use Areas, which will allow all commodity industry uses on an as-needed basis in times of high demand. The present Wilderness System would be redesignated with approximately 1 million acres of scattered Human Exclosures; 20 million acres of Wild Solitude Lands; 30 million acres of Backcountry; 30 million acres of Frontcountry; and 10 million acres of Commodity Use Areas.

Congress must also address the serious question of continuing to operate the National Wilderness Preservation System at a deficit. Vast amounts of natural resources are contained within Wilderness boundaries and substantial annual appropriations go to maintain hiking trails, camp sites, fire rings, horse trails, primitive toilets, and other facilities, and large amounts of taxpayer money go into studies of Wilderness, yet the Wilderness system has operated at a deficit every year since it was established in 1964. A Wilderness User fee must be established comparable to the entry fee program employed by the National Park Service. Wilderness is not a free good. It costs all taxpayers and benefits only a small minority. Only the affluent and well educated use Wilderness areas. Fewer than .01 percent of all Wilderness users consist of the educationally and economically disadvantaged. Wilderness users should pay for their recreation.

23. *Standing To Sue In Defense Of Industry.* Just as environmentalists won standing to sue on behalf of scenic, recreational and historic values in the 1965 case Scenic Hudson Preservation Conference v. Federal Power Commission, so pro-industry advocates should win standing to sue on behalf of industries threatened or harmed by environmentalists. Today, a specific individual or firm that is harmed must join a lawsuit as a plaintiff and pro-industry advocacy groups such as the Center for the Defense of Free Enterprise cannot

bring a lawsuit as a party of interest, despite their years of advocacy in support of business and industry. Because industries must continue to live with their regulators after lawsuits are settled they are hesitant to bring legal action in all but the most horrendous circumstances, which chills their access to justice. Just as Scenic Hudson conferred standing to sue on organizations devoted to saving natural features, recognizing them as harmed parties, so our court system must confer standing to sue on organizations devoted to saving industry, recognizing them as harmed parties. There is no symmetry today between the rights of environmentalists to sue and the rights of pro-industry advocates to sue. This is not fair and must be changed in the name of justice.

24. *National Recreation Trails Trust Fund.* Trail enthusiasts using motorized vehicles pay millions in federal gasoline taxes annually which are used to construct highways, not aid motorized recreation programs. These monies should instead be returned to a National Recreational Trails Trust Fund. The fund would provide matching grants to state and federal land management agencies, and local governments, coordinated through appropriate state agencies (state parks and recreation departments or departments of natural resources) with the primary goal of encouraging multiple-use trail development. A provision should also be made for adding additional revenues to the fund in the future, revenues derived from trail activities which do not generate fuel taxes.

25. *The End of the "Let Burn" Policy.* All naturally-caused wildfires in national park units and wilderness areas will be immediately and effectively extinguished to prevent the loss of natural and economic values. More importantly, all ground fuel accumulations which could lead to disastrous wildfires shall be actively prevented by a wise use management program.

Wildfire Prevention. All national park and wilderness areas will be managed to prevent the long-term buildup of ground fuels such as dead and down trees that create the ignition base for wildfires. Prevention must be actively pursued on all areas and is not optional. Prevention techniques may be of two kinds, to be permitted by temporary suspension of the Wilderness Act of 1964 in affected areas to allow motorized vehicles proper economic and rational access to danger sites:

(1) Managed Fuel-Reduction Burning. Fuel accumulations of dead and down wood in all national parks and wilderness areas will be periodically inspected, gathered and moved by tractor into appropriate batches and burned in accordance with state forestry regulations. Areas will be restored to pre-burn condition within two calendar years after managed fuel-reduction burning by hand raking crews and planting the affected area in fast-growing native indigenous herbaceous ground cover plants.

(2) Commercial Fuel-Reduction Harvest. Fuel accumulations of dead and down wood in all national parks and wilderness areas will be periodically inspected, gathered and chipped by motorized portable equipment, and the chipped wood removed from the natural area. Chip transport trucks shall take the chips to the nearest mill willing to buy the chips and sold at market prices. Areas will be restored to pre-chipping condition within two calendar years after commercial fuel-reduction harvest by hand raking crews and planting the affected area in fast-growing native indigenous herbaceous ground cover plants.

The neglect by any national park or wilderness administrator of ground fuel accumulations shall be a felony malfeasance of office subject to severe penalties of fine and imprisonment. A ground fuel wildfire, but not a crown fire, on any national park or wilderness area shall be prima facie evidence of negligence and malfeasance.

With such practical and beneficial techniques ready to hand there is no excuse for such disasters as the Yellowstone Holocaust of 1988.

Index

Abbey, Edward, 13, 209, 263
Acceptable risk: in Clean Air Act amendments, 87. *See also* Risk assessment
Acid rain, 110, 111, 113, 123, 135, 141
Ad hocracy, environmental, 126–40. *See also* Anti-toxics activism
Adams, Ansel, 20, 69–70
Adams, John, 74, 155, 179, 180, 237, 262; and NAFTA negotiations, 184, 186; at People of Color Summit, 152–53
Addams, Jane, 21
Aesthetics, 3, 20, 45, 127, 174
African-American children, and lead, 145, 155
African-American communities, 152; proximity to toxic-waste dumps, 141–42
African-American health, 141, 145–46
Agriculture: chemical-based, 72; Euro-American, 11, 14ff.; organic, 72
Agro-biotech industry, 137
Air pollution, 7, 14; and Clean Air Acts, 23, 33, 36, 81, 86–87, 88, 109, 117, 123, 136–37, 144, 159, 193, 243
Air quality: demographics, 144; regulations, 182; standards, 112, 144, 182; in workplace, 157–58, 171
Alabamians for a Clean Environment, 130
Alaska, 36, 71, 248–49
Alaska Lands Act (1980), 193
Allen, Robert, 68–72
Alliance for America, 85, 95
Alston, Dana, 151–52, 153, 154, 174, 262
Altgeld Gardens (Chicago), 141
Altman, Roger, 178
American Association of Fundraising Council, 40
American Barrick Resources, 183
American Greens, 242
American Littoral Society, 138
American Mining Congress, 98
American Petroleum Institute, 98
Ames, Amyas, 59
Ancient Forest Alliance, 213
Anderson, Bruce, 200–202
Anti-environment lobby, 66, 192–93
Anti-environmentalism, 64, 78, 83–103; mainstream response to, 102–103; violence and, 97, 98; in Reagan administration, 65ff. *See also* Sagebrush Rebellion, Wise Use movement
Anti-immigration opinion, 162–66
Anti-pollution industry, 63–64. *See also* Waste-management industry
Anti-toxics activists, 127–40, 207, 208, 237; effect on polluting industries, 133, 134, 145; and environmental justice movement, 169; fight against incinerators, 181; opposed to risk assessment, 122; people of color and, 144, 220–21; women in, 129ff.; and workplace, 159;
Anti-toxics environmentalism, early, 21–22;
ANWR. *See* Arctic National Wildlife Refuge
Appropriations, congressional, 194, 213
Arctic National Wildlife Refuge (ANWR), 45
Armstrong, Jeanette, 233–34
Arnold, Ron, 93–97

Arthur, William, 218
Asbestos ban, 79–80
ASCMEE. *See* Association of Sierra Club Members for Environmental Ethics
Asetoyer, Charon, 152
Ashland Principles, 231
Asian Americans: fish consumption of, 156; proximity to uncontrolled toxic-waste sites, 145
Association of Sierra Club Members for Environmental Ethics (ASCMEE), 216–18
Asthma, in urban black children, 145–46
Audubon, John James, 14, 20
Auto emissions, 68, 112, 182
Avila, Ausensio, 149
Ayres Richard, 71, 123

Babbitt, Bruce, 91, 177, 237; and Everglades cleanup, 183–84; and Surface Mining Control and Reclamation Act (SMCRA, 1977), 182–83; on timber sales, 190
Baca, Jim, 184
Backlash. *See* Anti-environmentalism
Bacon, Francis, 12–13
Bahouth, Peter, 74, 116
BASF plant, 159–60
Becker, Dan, 182, 193
Beltway environmentalists. *See* Mainstream nationals
Beltway strategy. *See* Federal strategy
Bensman, Jim, 217
Bentsen, Lloyd, 178
Berg, Peter, 229
Berle, Peter A. A., 57, 247
Berry, Thomas, 222, 255
Beyond the Green: Redefining and Diversifying the Environmental Movement, 154
Bhopal, 158
Big Four, and timber industry, 213
Big Green (California Proposition 139), 115
Billings, Leon, 81
Biodiversity Treaty, 191
Bioregionalism, 229–30
Biosphere, 13, 141, 234, 262
Biotechnology, 137
Biotic community, 19, 22
Birth control technologies, 162
Birth defects, 128, 259
Bison, 11, 14, 148
Black Caucus. *See* Congressional Black Caucus
Blue-collar enviros, 158–60; and mainstream nationals, 170
BLM. *See* Bureau of Land Management
Bob Marshall Wilderness Area, 252
Bookchin, Murray, 161, 224, 228–29, 239
Boone and Crockett Club, 15, 16
BP. *See* British Petroleum
Brandborg, Stewart, 25, 49, 61, 193
Bretton Woods, and third world development, 118
British Petroleum (BP), contributes to Earth Day programs, 27
Brookhaven (N.Y.) Town Natural Resources Coalition (BTNRC), 34
Brower, David, 1, 30, 31, 61, 208–209; and compromise as an environmental tactic, 76; co-founds League of Conservation voters, 75–76; founds Earth Island Institute, 75; founds Friends of the Earth, 75; and Glen Canyon dam, 76; and minorities in Sierra Club, 163–64
Brown, Ron, 178
Browner, Carol, 177, 237, 256; appeals to mainstream leaders, 198; and NAFTA negotiations, 185; sued by Clean Ocean Action, 138 and third-wave environmentalism, 181; and word *justice*, 157
Browning-Ferris, 64, 134
Brundtland, Gro Harlem, 235
Brunner, Calvin, 131
Bryant, Pat, 152
BTNRC. *See* Brookhaven Town Natural Resources Coalition
BTU tax, 183
Bullard, Robert, 147, 154, 170, 172, 252, 262
Buntrock, Dean, 55, 113–14
Bureau of Land Management (BLM), 184, 213; hostility toward, 91; lands claimed by Nevada statute, 90; and Mountain Ute grazing, 148; as subsidiary of extractive industries, 64
Burford, Anne (Gorsuch), 65, 67, 91–92
Burford, Robert, 91–92
Bush administration, 6; and NAFTA, 185
Bush, George, 27, 66, 68, 108, 117, 179, 181, 182, 196; and 1990 amendments to Clean Air Act, 87–88; appoints Reilly EPA administrator, 90; court appointments of, 79, 99; and Earth Summit, 167

Buzzelli, David, 237

CAFE standards, and GATT, 188. *See also* Corporate average fuel economy
California Coastal Commission, 99
California Coastal Conservancy, 74
California Forest and Watershed Council, 215
California Proposition 130 (Forests Forever), 115, 215
California Proposition 138 (Big Green), 115
California Rural Legal Assistance (CRLA), 150
California, strict air standards of, 182
California Waste Management Board, 142
Campaign finance reform, 194
Campana Amazonia por la Vida, 120; and NRDC compromise in Ecuador, 120
Cancer, 80, 121, 145, 152, 156, 159; pesticide-induced, 122
"Cancer Alley" (La.), 152
Cancer bonds. *See* Pollution credits
Capitalism, 28, 124, 223, 263
Carbon dioxide, third world emissions of, 119
Carbon monoxide, 21, 86
Carbon tetrachloride, 160
Carcinogens, 256; in food, 137, 181. *See also* Delaney amendment
Carson, Rachel, 1, 12, 21, 23, 24, 35, 85, 106, 178, 224, 225, 263
Carter. Jimmy, 26, 89, 128, 171
Cattle industry, 85, 92
CCHW. See Citizens' Clearing House for Hazardous Wastes
CDFE. *See* Center for the Defense of Free Enterprise
Center for Community Economic Research, 240
Center for Policy Alternatives, 57–58
Center for Science and Environment (New Delhi), 119
Center for the Biology of Natural Systems, Queens College, CCNY, 40
Center for the Defense of Free Enterprise (CDFE), 93
CEQ. *See* Council on Environmental Quality
Cerrell Report, 142, 149
CF. *See* Conservation Foundation
CFCs. *See* Chlorofluorocarbons
Chavez, Cesar, as *environmentalist*, 172
Chavis, Benjamin, 141, 237
Chemical Manufacturers Association PACs, 193
Chemical pollution, 7, 21
Chem Waste: Alabama hazardous-waste dump of, 130; Kettleman City incinerator, 149–50
Chemical Waste Management Corporation (WMX) *See* Chem Waste
Chevron: environmental philanthropy of, 54; Papua New Guinea oil pipeline, 119
Children's Defense Fund, 74
Chlorofluorocarbons (CFCs), 184
Christian Right, 84, 232
Chrysler strike, 159
Citizens party, 242
Citizens Trade Campaign, and NAFTA, 187
Citizens' Clearing House for Hazardous Wastes (CCHW), 147, 208, 237; founded, 128; and Green Group, 74; and McDonald's campaign, 139; and people of color, 144; scope of outreach, 133. *See also* Lois Marie Gibbs
Citizens' Committee for the Right to Keep and Bear Arms, 93
Citizens' Laboratory, 132
Citizens-for-clear-air groups, 81
Civil rights environmentalism, 8, 168–69, 171
Civil rights leaders: and environmental racism, 144; and mainstream's racist hiring practices, 146
Civil rights movement, 252; and Earth Day 1970, 25
Civil rights, 25, 252; and Clinton order on environmental equity, 191; nexus with environmentalism, 8, 168–69, 171
Clark, Mike, 74
Clean Air Act (1967, 1970), 23, 33, 243; inequitable enforcement of, 144; lawsuits to enforce, 36, 144; national air ambiance standards in, 81; reduction goals of, 86; weakening of, 86–87
Clean Air Act (1990), 193; accommodations to industry, 136–37; coal miners protest, 159; creates pollution credits, 87, 88, 109ff., 123; effect on pollution, 117
Clean Air Coalition, 87
Clean Ocean Action (COA), and ocean dumping, 137–30
Clean Water Act: (1965), 23; (1977), 33
Clean Water Action Project, 132

Clearcutting, 48, 190, 230; on public lands, 213, 216
Clinton administration 177–92, 256; accepts cost-benefit analysis, 56; and Earth Day 1995, 202–203; environmental breakfast, 198–99 of; environmentalists in, 177–80; and NAFTA, 184–88; threatens to exempt timber industry from environmental laws, 190
Clinton, Bill, 100; and air quality standards, 182; and old-growth forests, 189; appoints environmentalists, 177–80; as Arkansas governor, 177; and campaign promises, 179; as "environmental president," 66, 194, 195, 203; executive order on environmental racism, 157; holds Forest Summit, 189; and GATT, 188; and mainstream environmentalists, 177ff.; promise on U.S. emissions, 181
Closing Circle, 23
Club of Rome, 161
COA. *See* Clean Ocean Action
Coal industry, 79, 183
Coal, low-sulfur, 113
Coal miners, 113, 159
Coalition for Clean Air in Los Angeles, 105
Coast Range Association, 103
Cockscomb Jaguar Preserve, 54
Colby, Michael, 206
Cole, Thomas, 20
Colorado squaw fish, 148
Columbia River Intertribal Fish Commission, 237
Commission for Racial Justice (UCC), 141–42, 155, 157
Commoner, Barry, 20, 23, 161; on banning chemicals, 40; concept of ecosphere and technosphere, 225; on impact of environmental movement, 6
Community: in Native American culture, 233–34; rights of, 135, 137
Competitive Enterprise Institute, 99
Competitiveness Council, 191
Comprehensive Environmental Response Compensation and Liability Act. *See* Superfund
Compromise, as environmental tactic, 6, 52, 59–60, 66, 72, 76–77, 87, 217, 218, 259; and anti-toxics activists, 133; with carcinogens, 122; and green corporate images, 140; institutionalized in ERCs, 110; and mainstream nationals, 129, 173; as third-wave strategy, 107, 114; in third world environments, 121
Concerned Citizens in Action, 130, 139
Congressional Black Caucus, 154; and NAFTA, 173; LCV rating of, 172
Conoco: oil drilling in Amazon, 119
Conservation Exchange newsletter, 116
Conservation Foundation (CF): conducts Superfund study, 89–90; and corporate philanthropy, 115; established, 55; supports Earth Day 1970, 25–26
Conservation Fund, 171
Conservation International, supports NAFTA, 186
Conservation movement, 1–2, 8, 10, 106; absorbed by environmental movement, 23; accepts Pinchot's ideas, 17; becomes environmental movement; elitism of, 2–3, 151. *See also* Environmental movement
Conservationism: as defined by Pinchot, 16–18, 19; as dominant paradigm of federal land management, 17–18; conflict with preservationism, 16–18, 68
Consumer-based economies, 188, 255
Convention on Climate Change, 181
Cook, John, 155
Coors, Joseph, 67, 84, 92
Corporate average fuel economy (CAFE) standards, Clinton and, 182
Corporate compliance lawyers, 38, 78
Corporate Conservation Council, 115
Corporate lobby, 211: growth of, 85; resources of, 86, 192–93; strategy of, 86, 88
Corporate Philanthropy Report, 55, 58
Corporate philanthropy: 53–59, 253; effect of *Exxon Valdez* spill on, 54; for environmental seminars, 54–55; to improve environmental reputation, 54, 115
Corporate polluters, lawsuits against, 35–37
Corporate public relations, 117, 250
Cost-benefit analysis, 18, 56, 108; in asbestos case, 80; Clinton administration discards for environment; 191; and costs of preventing death, 122; and environmental regulations, 67, 191; and EPA, 122; and federal courts, 79–80. *See also* Risk assessment
Council on Economic Priorities, 237
Council on Environmental Quality (CEQ),

32–33, 178; budget cut, 67; *Global 2000 Report* of, 65
Counter-environmentalism. *See* Anti-environmentalism
Cowboy Caucus (Utah), 101
Cox, Robert, 169
Coyle, Marcia 143–44
Craver, Matthews and Smith, 45, 47, 70
Craver, Roger, 45, 47, 176
CRLA. *See* California Rural Legal Assistance
Crockett, Kate, 136
Crook, George, 148
Culture of reform, 7, 29–61; foundation influence on, 49, 52
Cuyahoga River fire, 24, 128

Daley, Bill, 186
Darling, Jay "Ding," 19
DBCP, worker exposure to, 158
DDT ban, 34–35, 39
Debovoise, Dickinson, 138
Deep ecology, 226–28, 235
Defenders Network, 99
Defenders of Property Rights, 98
Defenders of Wildlife, 187, 193; Hair vetoes G-10 membership, 69; growth in Reagan years, 70
Delaney amendment, 137, 181, 256. *See also* Carcinogens, Cancer
Democracy: in environmentalism, 206, 243; in grassroots movement, 135
Descartes, René, 12, 13
Devall, Bill, 224, 227
Die Grünen, 241,
Dinosaur National Park, 76
Dioxins, 128; EPA limits in fish, 156; in Passaic River bottom, 138
Direct action, 46–47, 77, 240, 252
Direct mail, 60, 61, 240; declining effectiveness, 176; campaigns against Reagan administration, 68; as education, 44–45; as environmentally degrading, 43; as funding source, 41, 44; dependency, 47–49; expense, 43; influence on agenda, 41; limitations of, 102; list fatigue in, 46; loss of effectiveness, 44, 55; membership recruitment by, 42; overexploitation of issues, 47; and third-wave environmentalism, 117
Direct mail managers: influence of, 44, 45
"Dirty Dozen," 26
Discourse on Inequality, 12

Diversity, in environmental movement, 146–47, 262
Dolan v. *Tigard* (Ore.), 99–100
Dole amendment, 101
Doniger, David, 80
Dougherty, James, 193
Dow Chemical, 89, 132
Dubos, René, 259
Dudek, Dan, invents ERCs, 109
Dudley, Barbara, 74
Duggan, Sharon, 78, 150
Duggan, Stephen, 35
Dunlap, Riley, 23, 205–206
DuPont, 89, 184

Earth Day 1970, 3, 5, 23, 63, 24–25, 200, 239
Earth Day 1980, 26
Earth Day 1990, 26, 175, 230
Earth Day 1993, 192
Earth Day 1994, 192
Earth Day 1995, 27, 133, 200–203
Earth First!, 13, 209–210
Earth Force, 52
Earth Island Institute, 75, 209
Earth Summit, 166–67, 182, 191
ECO. *See* Environmental Careers Organization
Ecofeminism, 231
Ecological crisis, 231, 259
Ecological economics, 235, 245–46
Ecological ideology: need for unifying, 221–26, 262; as synthesis, 255–57
Ecology movement, 16, 23, 28
Ecosocialism, 240
"Ecotage," 209
Ecuador, oil drilling in, 119–20
Ehrlich, Paul, 23, 161
El Pueblo para Aire y Agua Limpio, 149–50
Electrical utility industry, 182, 183
Emission reduction credits (ERCs). *See* Pollution credits
Emissions, automobile, 112, 124, 182
Endangered species, 191, 195, 251, 262
Endangered Species Act (1973), 33, 37, 177, 190; targeted by anti-environmentalists, 200, 255
Engels, Friedrich, 239
Enviro-tech companies. *See* Anti-pollution industry
Environment, as word, 1, 5–6, 21, 23, 34
Environment: as limitless, 206; of workplace, 157–60

Environment-civil rights nexus, 168, 171
Environment-human rights nexus, 208, 245
Environmental Action, 26
Environmental ad hocracy, 126–40, 208. *See also* Anti-toxics activism
Environmental Agenda for the Future (1986), 71–72
Environmental backlash. *See* Anti-environmental movement, Sagebrush Rebellion, Wise Use movement
Environmental Careers Organization (ECO), employment survey of, 154
Environmental cleanup industry. *See* Anti-pollution industry
Environmental Coalition for NAFTA, 186, 187
Environmental Conservation Organization, 96
Environmental Defense Fund (EDF), 23, 198, 199, 238; and Clean Air Act (1990), 87, 117; and COA lawsuit, 138–39; corporate executives on board of, 58; divided over NAFTA, 187; and ERCs, 109; and FAIR, 163; fires Yannacone, 38; founded, 35; and McDonald's packaging, 139–40; and North Carolina biotechnology law, 137; screens corporate donors, 115; and "Stockholm Syndrome," 116
Environmental disasters, 42, 128–29; Bhopal, 158; Cuyahoga River fire, 24; *Exxon Valdez* oil spill, 54; Hanford nuclear facility, 158; Love canal, 127–28; Santa Barbara oil spill, 24; Three-Mile Island, 129
Environmental-economic tradeoff, 158, 160, 192
Environmental education, 44–45, 50, 52
Environmental enforcement, 64, 66; costs of, 81–82; via litigation, 78
Environmental equity, 156–57, 207
Environmental ethics, 11, 13, 19, 60, 162, 262
Environmental Grantmakers Association, 113
Environmental historians, 5, 19, 21, 24
Environmental ideology, 221–26, 255–57
Environmental imagination, 7–8, 9–28, 20, 27, 20, 27, 125, 206, 208, 224, 239, 259, 260
Environmental impact statements, 32, 33
Environmental injustice, 20, 144
Environmental justice, 8, 47, 191, 220, 226, 240, 252, 261, 262
Environmental justice movement, 125–74; and anti-toxics activists, 169; democratic decision making in, 154; difficulty of fundraising for, 153; as international issue, 166; and mainstream nationals, 168; and People of Color Leadership Summit, 151ff.; and workplace environment, 158–60
Environmental law, 27, 38
Environmental law firms, 35–38. *See also* Environmental Defense Fund, Natural Resources Defense Council, Sierra Club Legal Defense Fund
Environmental Law Center, 38
Environmental Law Institute, 55
Environmental legislation, 5, 7, 8, 33, 34; backlash against, 181; costs of, 81–82; endangered by takings initiative, 100–101; effect of GATT on, 188; effect of NAFTA on, 188; enforcement of, 36–37, 64, 66, 188, 213; in 1994 Congress, 199; reauthorizations of, 81, 192, 194, 200
Environmental lobby, 33, 180; access to legislators, 85, 218–19; compared to corporate lobby, 86; financial resources of, 192; growth in Reagan years, 72; and power of green vote, 195; weakness of, 192
Environmental movement, 10; beginning of, 23; and civil rights, 8, 168, 171; and community activism, 240; definition of, 26; development of, 27–28; impact of, 6–7; mainstream, 3–8; need for grassroots strategy, 64–65, 102–103; need for realignment, 173; and nonhuman life, 255; as social movement, 2–3, 8, 10, 263; as white men's club, 8, 219–20. *See also* Anti-toxics activism, Conservation movement, Environmental justice movement, Grassroots activism, Mainstream nationals
Environmental Opinion Studies (EOS), 43, 195–96
Environmental philanthropy, 236; disproportionate distribution of, 248, 250; effect of shrinkage in, 248–50; influence on agenda, 41; sources of, 40–59. *See also* Philanthropy
Environmental populism, 129ff. *See also* Anti-toxics activism
Environmental Protection Administration

(EPA), 6, 31, 108, 243; asbestos ban overturned, 79–80; avoidance of word *justice*, 156; bans leaded gasoline, 155; Browner appointed, 177; Anne (Gorsuch) Burford appointed, 67; and cabinet status, 101; charges of racism against, 156; COA suit on ocean dumping, 138; created, 32; and Delaney amendment, 256; and dioxins in fish, 156; embraces cost-benefit analysis, 122; emissions limits and, 109; environmental equity report, 156; and interests of minority communities, 155; and Kettleman City, 149–50; at Love Canal, 128; memo on environmental justice, 157; Pollution Prevention Policy Statement, 117; regulation of harmful substances, 60; and Superfund study, 89–90; and third-wave environmentalism, 181; Toxic Release Inventory, waste-disposal regulations, 114; and waste reduction, 145; 108; and WMX incinerator, 113–14
Environmental racism, 143, 144, 156
Environmental regulations: public support for, 72–73; as takings, 98
Environmental science, 27, 39–40, 108, 119
Environmental vote. *See* Green vote, Voting
Environmental-economic tradeoff, 73, 192, 196
Environmentally Conscious Manufacturing, 107
EOS. *See* Environmental Opinion Study
EPA. *See* Environmental Protection Administration
ERCs. *See* Pollution credits
Espy, Michael, 237
Ethics, environmental, 11, 13, 19, 60, 162, 216–18, 226, 262
European Green parties, 228, 241–42
European settlers, 10–11, 14 ff.
Evans, Brock, 94, 179, 189
Everglades cleanup costs, 183–84
Evergreen Foundation, 96
Evers, Medgar, 7, 169, 252
Everyone's Backyard, 133
Extractive industries, 64, 225; support anti-environmental movement, 84, 85, 95, 103
Exxon Valdez, 54, 57

FAIR. *See* Federation for American Immigration Reform
FAIR resolution, 164
Family planning, 162, 191
Fanjul, Alonso, 184
Farm workers: exposure to pesticides, 155: cancer cluster, 156; health problems, 145
Fauntroy, Walter, 142
Federal air quality standards, 81, 112, 182
Federal courts, 36–37; changes in, 78; and environmental regulation, 79; and cost-benefit analysis, 79; Reagan and Bush appointments, 79
Federal government, as environmental scofflaw, 37
Federal Insecticide, Fungicide, and Rodenticide Act (1972), 33
Federal strategy, 64, 65ff., 173, 176, 188; and stricter state standards, 81. *See also* Legislative/litigative strategy
Federation for American Immigration Reform (FAIR), 163, 164
Feldman, Jay, 136
Feminist ecology, 230–31
Ferris, Deeohn, 155
Fifth Amendment, 98, 100
First National People of Color Leadership Summit. *See* People of Color Leadership Summit
Fischer, Michael, 147, 218–19; on mainstream hiring, 147; and G-10, 70, 74; and Sierra Club minorities, 164; at People of Leadership Color Summit, 152
Fishing industry, 183, 188
Flo-Sun Corporation, 184
FOE. *See* Friends of the Earth
Foley, Thomas, 183
Ford Foundation, 247; funds environmental law firms, 35–36, 37; litigation review committee, 38; reins in environmental lawyers, 37–38
Foreman, David, 166, 208–211, 212, 222, 223, 224, 229, 256
Forest activists, 103, 189–90, 212, 231, 261
Forest management, and foundation influence, 52
Forest preservation, 208
Forest Summit, 189–91
Forests Forever (California Proposition 130), 115, 215
Foundations, 49–53, 253: critique of Green Group, 50; and environmental law firms, 35–38; environmental programs and strategies, 49–53; and

grassroots organizations, 250; oversight of spending, 49; and population issues, 163, 165, 166; reformist influence of, 49; support of mainstream nationals, 41, 49. *See also* Philanthropy
Fourth-wave environmentalism, 8, 215, 221, 231, 234, 238, 255; diversity of, 207–208, 257; ideologues of, 224ff.; and new civil authority, 221, 246
Frampton, George, 179
France, Tom, 189
Frank Weeden Foundation, 165
Frank, Judy Knight, 148
Free market ideology, 91, 98, 246; environmentalist accommodation to, 72
Free trade, and third world ecosystems, 118
Friends of La Plata, 148
Friends of the Earth (FOE), 45, 186, 187, 209; and Green Group, 74; growth in Reagan years, 70
Frome, Michael, 253
"Fugitive emissions," in GM plant, 159
Fundacion Rio Napo, 120

G-10. *See* Group of 10, Green Group
Gaia hypothesis, 14
GAO. *See* U.S. General Accounting Office
Gasoline lead emissions, 68, 112
GATT. *See* General Agreement on Trade and Tariffs
General Agreement on Trade and Tariffs (GATT), 188
General Electric, 89
General Motors, 124, 159
Genetic engineering, 137
Genocide, 152, 162, 166
Georgia Pacific, 215, 228
German Greens. *See* Die Grünen
Gibbs, Lois Marie, 128–30, 132, 133, 135, 136, 144, 153, 171–72, 214, 221, 2237, 254; and direct action, 129; as *environmentalist*, 172; and McDonald's campaign, 139; as model for women activists, 130; organizes Love Canal Homeowners Assn., 128; starts Citizens' Clearing House for Hazardous Wastes, 128;
Gibbs, Michael, 128
Gillenkirk, Jeffrey, 42, 48
Glen Canyon dam, 76
Global 2000 Report to the President, 65
Global environment, 119, 162, 165, 236
Global Stewardship grants, 165
Global warming, 71, 119, 135, 167, 181, 236, 250
Gore, Albert, 66, 177ff., 184, 190, 192; and campaign promises, 179; and Environmental Justice Act, 154; environment-health nexus, 198; and Liverpool, Ohio, incinerator, 181
Gottlieb, Robert, 21, 252
Gottlieb, Alan M., 93
Governmental Refuse Collection and Disposal Association, 131
Grand Canyon, 20, 30–31, 126, 158, 171
Grant, Madison, 2
Grassroots activists, 5, 8, 126–40; 211, 256, 261–62; compete for funding scraps, 248; criticisms of mainstreams, 135, 136–39; criticism of G-10 *Agenda*, 72; distrust of federal strategy, 65; excluded from forest discussions, 52; and foundation influence, 52; friction with mainstreams, 214, 215–16; and Option 9, 189–90; need for mainstream cooperation, 147; response to Clinton administration, 179; and risk assessment, 121; tactics, 132; and third-wave environmentalists, 113; and traditional conservation issues, 135; view of mainstream boards, 58; view of federal strategy, 59; and workers, 159; *See also* Anti-toxics activists, Environmental justice, Forest activists
Grassroots network on oil and gas wastes, 58
Grassroots organizing, of Wise Use, 102
Grazing fees, 90, 92, 95, 184, 195
Grazing rights of Mountain Utes, 148
Great Lakes, mercury in, 39
Great Old Broads for Wilderness, 211
"Green capitalist," 239–40
"Green fire," 18–19
"Green funds," 64
Green Group, 50, 198, 214, 254; effectiveness of, 74; expansion of, 74; focus on winning, 171–72; lobbying budgets of, 85; WRI joins, 119. *See also* Group of 10
Green parties, 224, 241–42
Green vote, 195–98
Greenhouse gases, 109, 111, 119
Greenpeace, 116, 211, 247, 254; alliances with local activists, 132, 147; direct mail and, 46–47; excluded from G-10, 69; excluded from G-10, 69; joins Green

Group, 74; and Kettleman City, 149; membership declines, 46;
Greenwashing, 27, 201, 250
Greer, Sue, 130
Greider, William, 79, 170, 173
Griffin, Melanie, 110
Grinnell, George Bird, 14
Group of 10 (G-10): established, 68–69; expanded, 74; media see as environmental movement, 70; original members, 69; renamed Green Group, 74. *See also* Green Group
Gulf Coast Tenant Leadership Development Project, 146
Gulf Coast Tenants Association, 152

Hair, Jay Dee, 11, 107, 237; and corporate ideology, 116; invents Corporate Conservation Council, 115; and Green Group, 68, 69, 74; mediates EPA-WMX waste-disposal issue, 113–14; and NAFTA negotiations, 184, 185, 187
Hamilton, Alice, 21–22, 263
Hanford, Wash., nuclear facilty, 158
Hardin, Garrett, 162, 166
Harp seals, 42, 46
Hatch, Orrin, 91
Hatfield, Mark, 214
Hawkins, Howard, 223
Hayduke, George Washington, 209
Hayes, Denis, 24–26, 27, 200, 201–202
Hayes, Randy, 194
Hays, Donald, 35
Hays, Samuel, 24
Hazardous-waste facilities: in nonwhite communities, 251
Hazardous-waste facilities: and Environmental Justice Act, 154; proximity to nonwhite communities, 141, 142; opposition to, 130, 142, 143, 145; siting of, 134, 141–45, 149
Hazardous-waste industry, 133, 134, 135
HCFCs, 184
Health problems: of Love Canal children, 128; of minorities, 145–46, 155–56
Health-environment nexus, 198–99
Heartwood, 52, 211, 213
Heinz, John, 107
Helvarg, David, 94, 95
Heritage Foundation, 68, 84, 99, 107
Hermack,Tim, 52, 189–90, 211
Hetch Hetchy dam, 17
Hickel, Walter, 177
Hickman, H. Lanier, Jr., 131
Hill and Knowlton, 53
Hills, Carla, 185, 188
Hinds, Cathy, 130
Hispanic Americans: at Kettleman City, 149–50; pesticide-related illness of, 145, 156; proximity to toxic-waste disposal sites, 142, 144–45
Hodel, Donald, 67, 177
Holmes, William, 97
Hooker Chemical Company, 127
Hopewell, Va., poisoned wells, 128
Hornaday, William, 2
Howe, Sydney, 26
Huaorani people, 119
Human consumption, 26
Human health, 255; and environmentalism, 222
Human right, clean environment as, 33, 226, 244–45
Human Rights Watch, 244, 245
Humanity-in-Nature, 224, 233
Hume, David, 12
Hutton, James, 13
Hyde, Phil, 20
Hydroelectric power, 183

Ideological diversity, 255, 262
IMF. *See* International Monetary Fund
Immigration, 162–66
Incinerators, 131, 135; in Liverpool, Ohio, 179, 181; near nonwhite communities, 145, 149–50
Indian reservations: natural resources on, 220; nuclear testing and storage on, 220; strip mining on, 183; waste-disposal on, 220
Industrial lobby. *See Corporate lobby*
Industrial economies, 166, 239, 255
Industrial polluters, 261; and Clinton administration, 183, 184, 185
Industrial waste reduction, goal of NIMBYs, 133, 135
International environmental movement: and third wave, 118–21
International Monetary Fund (IMF), 118

Jackson, Leroy, 97, 172
Jesse Smith Noyes Foundation, 236
John Muir Society, 217
Johnson, Hazel, 130
Johnson, Lyndon, 31, 32
Johnson, Robert Underwood, 17

Judaeo-Christian culture, 10–11, 232, 255
Judicial review, of timber sales, 190
Justice, environmental, 125–74

Kalaw, Maximo 167
Kantor, Mickey, 185ff., 188
Kellett, Mike, 211
Kendall Foundation, 68
Kennard, Byron, 26
Kennedy, Robert, Jr., 119–20
Kepone poisoning, of Hopewell, Va., wells, 128
Kettleman City, Calif.; citizen opposition to Waste Chem incinerator, 149–50
Kerr, Andy, 103
Kidder, Peabody, 59
Kiker, Kaye, 130
King County (Calif.) Board of Supervisors, sued by El Pueblo para Aire y Agua Limpio, 150
King, Martin Luther, Jr., 140, 252–54, 263
Klaes, Michel, 169
Krupp, Fred, 48, 58, 124, 184, 187, 198, 222, 237, 254; on Clinton, 182; and corporate ideology, 116; and McDonald's, 116, 139–40; promotes market incentives, 108

La Force, Norman, 165
Labor Community Strategy Center, 160, 243
Labor movement, 127
Labor environmentalism, 21–22, 159–60
Land ethic, 10–11, 18–19, 226
Land stewardship: conflict over, 11; of early naturalist writers, 14; Judaeo-Christian concept of, 10; Native American concept of, 11
Land-Air-Water-Law Conference, 150
Landfills, 113, 135
Landor, Walter, 247
Latham, Watkins, 105–106
Lathrop, Calif., workers and sterilizing chemicals, 158
Lavelle, Mariane, 143–44
Lazare, Daniel, 176
LCV. *See* League of Conservation Voters
Lead: airborne, 39; in bloodstream, 33, 145, 155; gasoline emissions reduced, 36; poisoning, 7, 21, 22
Leaded gasoline, EPA bans, 155
League of Conservation Voters (LCV), 43, 44, 46, 76, 172, 177, 182, 195, 198, 209; grades Clinton C+, 191
Lee, Charles, 155, 157
Left Green Network, 242
Legal Environmental Assistance Fund (LEAF), 35
Legislation-litigation-lobbying (Three Ls), 173
Legislative/litigative strategy, 6, 8, 65, 72. *See also* Federal strategy
Legislative strategy, 34, 59–60, 78; grassroots criticism of, 72; limitations of, 88; for reauthorizations, 200
Leopold, Aldo, 12, 18–19, 30, 60, 224, 227, 256, 262, 263,
Levitas, Steve, 137
Lewis, John, 172
Li, Vivian, 164
Limbuagh, Rush, 96, 102, 103
Litigation, environmental: implicit threat of, 114–15
Litigative strategy, 34–38, 78–80, 256; successes of, 36–37; anti-environmental response to, 83; as environmental enforcement, 78
Liverpool, Ohio, incinerator, 47, 179, 181
Lobbying: as environmental tactic, 33, 59–61, 68; ineffectiveness of environmental, 192–94
Local clean air activism, and federal strategy, 81
Locke, John, 12, 13
Logging, in public forests, 217
Los Angeles Air Quality Monitoring Board, 243
Los Angeles Clean Air Coalition, 243
Los Angeles Regional Clean Air Incentives Program (RECLAIM), 111
Los Angeles air quality, 87
Love Canal, 74, 127–29, 131, 155
Loveladies Harbor v. *United States*, 99
Lovelock, James, 14
Low-sulfur coal, 220
Lucas v. *South Carolina*, 99

Maddy, Jim, 191
Magnuson Act, 188
Mahler, Andy, 213
Mainstream environmental organizations. *See* Mainstream nationals
Mainstream environmentalists: commitment to cause, 255–56; in Clinton administration, 178–79; form G-10, 69;

lack of vision of, 256; loyalty to Clinton administration, 188, 190–91; and President's Council on Environmental Cooperation, 184; rivalry among, 69–70; support NAFTA, 185–86; and White House breakfast, 198–99
Mainstream nationals, 227; absence from environmental justice struggles, 152, 162; acceptance of Option 9, 189, 190; achievements of, 32–34; accommodationism, 135; anti-immigration opinion and, 162–63; anti-toxics activists' views of, 129; blue-collar enviros and, 158; on Bush at Earth Summit, 167; claim credit for grassroots successes, 136, 139–40; community rights and, 137; compromise and, 6, 129; contributions to, 41; corporate representation on boards, 56, 58, 64; declining memberships, 4ff., 175–76; Delaney amendment and, 137; direct mail dependency, 42–49; and Earth Day 1970, 25; environmental justice and, 144, 153, 168; and federal strategy, 5, 64, 71, 73, 176, 194; and forest activists, 212–213; growth in 1980s, 70–71, 72; influence on global development, 237; internal splintering, 214; legislative strategy, 88; and Native Americans, 220–21; need for partnership with grassroots, 170; need to share resources, 170; nonwhite staff, 153, 154; at People of Color Summit, 152–53; organizations comprising, 5; opportunity for diverse movement and, 129; philanthropy and, 248; priorities, 32; racism, 254; reform agenda, 30, 64; response to Reagan administration, 67–73; response to Sagebrush Rebellion, 92; response to Wise Use movement, 102–103; restructuring, 247–49; and revenue losses, 248; and risk assessment, 121, 122; rivalry among, 70; splinters from, 208–212; staff growth, 61; trend to bureaucracy, 61; try grassroots organizing, 249; as white men's club, 8, 31, 73, 219–20; wilderness preservation and, 127
Malthus, Robert, 161
Mandate for Leadership, 107
Mankiewicz, Frank, 53, 54
Mann, Eric, 160
Mares, Mary Lou, 149
Marine Mammal Protection Act (1972), 33, 188
Market incentives, 181, 106, 123, 106, 181; logic of, 108–109; problems with, 112
Marshall, Robert, 3, 251–52, 263
Marshall, Thurgood, 150–51, 253
Marston, Ed, 72
Marxism, 223, 239, 240, 255
Marzulla, Nancie, 98–99
Marzulla, Roger, 98–99
Mattole Valley, Calif., 230
Maya, Esperanza, 149, 150
McClarty, Mike, 178
McCloskey, Michael, 29, 116–17, 186, 187, 261
McCown, John, 169, 170
McDonald's, 116, 139–40
McFarland, Calif., cancer cluster in, 156
McGinty, Katy, 178
McGuire, Wally, 202
McKinley, William, 16
McMurray, David, 54
Meadows, Donnella, 161
Medical wastes, on beaches, 71, 260
Mendes, Chico, 97, 244
Merchant, Carolyn, 224
Mercury contamination, 7: exported to South Africa, 147; in Great Lakes, 39
Mertig, Angela, 23, 205–206
Methyl bromide, 185
Meyerhoff, Albert, 38, 79, 256
Military Toxics Campaign, 130
Miller, George, 197, 199
Mills, Stephanie, 214, 227, 229
Mining, 85, 90, 95, 184
Minorities, in mainstream organizations, 31, 146–47, 153, 154. *See also* People of color
Monkey Wrench Gang, 209
Monsanto, 89, 201
Montana Wilderness Bill, 180
Monteith, James, 193–94, 213
Moore, Curtis, 86
Moore, Richard, 170
More, Thomas, 12
Mountain States Legal Foundation, 92
Mountain Utes, 148, 221
Muir, John, 1, 11, 12, 170, 216, 224, 256, 263; conflict with Pinchot, 16–18; ethics of, 60; founds Sierra Club, 17; preservationism and, 16–17, 30; and wilderness, 16, 18
Multiple Use Strategy Conference, 93

Multiple use federal policy, 64

N.C. Department of Agriculture, 137
N.C. Genetically Engineered Organisms Act, 137
NAAEC. *See* North American Agreement on Environmental Cooperation
Naess, Arne, 226
NAFTA War Room, 186
NAFTA environmental side agreement. *See* North American Agreement on Environmental Cooperation (NAAEC)
NAFTA. *See* North American Free Trade Agreement
National Academy of Sciences, 122
National air ambiance standards, 81
National Association for the Advancement of Colored People (NAACP), 237
National Association of Realtors, 98
National Audubon Society, 179, 214, 215; in Ancient Forest Alliance, 213; attempts makeover, 247; board of directors, 55, 56–57; in Clean Air Coalition, 87; and corporate philanthropy, 115; declines to open legal department, 35; and Earth Day 1970, 25; and FAIR, 163; founded, 15; grant from WMX, 55; leadership supports NAFTA, 186; and old-growth forests, 52; oil and gas wastes project, 58; and population growth, 161–62; regional chapters oppose NAFTA, 187; uprising in, 219
National Campaign Against Toxic Hazards. *See* National Toxics Campaign Fund (NTCF)
National Coalition Against Misuse of Pesticides (NCAMP), 132, 136
National Economic Council, 178, 192
National Environmental Policy Act (NEPA, 1970), 24, 32, 178
National Forest Council, 217
National forests, 52, 161
National Highway Traffic Safety Administration, 182
National parks, 2, 17, 179
National Parks Service, 91
National Priority list (Superfund), 143
National Research Council, 145
National Toxics Campaign Fund (NTCF), 132, 147
National Wetlands Coalition, 85, 96
National Wilderness Preservation Sysem, 31, 33
National Wildlife Federation (NWF), 19, 68, 69, 214, 237, 254; budgets growth in Reagan years, 71; budget problems of, 175; in Clean Air Coalition, 87; Corporate Conservation Council of, 115; and corporate philanthropy, 54, 115; and Earth Day 1970, 25; grant proposal on environmental justice, 153; grant from WMX, 55; supports NAFTA, 185, 186; and old-growth forests, 52; and population issue, 152; and "Stockholm Syndrome," 116; Toxic 500 list, 56; WMX CEO on board of, 114
Native Americans, 145, 172, 220–21, 224; bioregionalism of, 12; community and, 233; conflict with European settlers, 148; fish consumption of, 156; displaced from Yellowstone Park, 12; driven onto reservations, 14; ecology of, 11, 233–34; grazing rights of, 148; land stewardship of, 11. *See also* Mountain Utes, Navajo, Ojibway, South Dakota Sioux
Native Forest Council, 52, 211
Natural Resource Cluster, 178
Natural Resources Defense Council (NRDC), 71, 74, 80, 105,179, 180, 182, 237, 247, 250, 256; in Clean Air Coalition, 87; divided over NAFTA, 187–88; enforcement lawsuits for Clean Air Act, 36; founded, 35; field offices of, 132; and industrial pollution, 162; joint project with Human Rights Watch, 245; Kettleman City and, 150; leadership supports NAFTA, 186; mediates oil drilling in Amazon rain forest, 119–21
Natural resources: conservation of, 126; exploitation of, 16; as inexhaustible, 14, 206; overconsumption of, 165
Nature Conservancy, 184, 249
Nature: desacralization of, in Enlightenment thought, 12–13; humanity's relation to, 224, 225; images of, 20
Navajo teenagers, cancer in, 145
NCAMP. *See* National Coalition Against Misuse of Pesticides
Negligible risk, 121–22, 181. *See also* Risk assessment
Nelson, Gaylord, 25, 162, 200–202
NEPA. *See* National Environmental Policy Act (1970)
Network for Environmental and Economic Justice, 147
Nevada initiative, 90–91

New Age Greens, 242
New civil authority, 221, 246
New Jersey beaches, medical wastes on, 71
New Left, 239–40
New world economic order, 168
New York State Health Commission, 128
New York Zoological Society, 2, 55
Newman, Nathan, 240
Newman, Penny, 130, 139, 140, 159
Newsom, Bill, 58
Newton, Isaac, 12, 13
Newtonian-Cartesian view, 255
NIABY (Not-In-Anybody's-Backyard), 133
Nichols, Mary, 179, 182
NIMBYs (Not-In-My-Backyard), 131, 133, 145. *See also* Anti-toxics activists
Nitrogen oxide, 111. *See also* Greenhouse gases
Nixon administration, 35
Nixon, Richard, 32, 178
Nollan v. *California*, 99
Nonwhite communities: population-immigration issue and, 166; proximity to toxic-waste sites, 141–42, 145
Noonan, Patrick, 171
NOPE (Not-On-Planet Earth), 135
Noranda Mines (Montana), 38
North American Agreement on Environmental Cooperation (NAAEC), 185, 188
North American Commission on the Environment, 186
North American Free Trade Agreement (NAFTA), 74, 173, 184–88; environmental side agreement, 185, 188
Northern Lights Institute, 92
NTCF. *See* National Toxics Campaign Fund
Nuclear energy, 71, 85, 148, 171
Nuclear exposure, 129
Nuclear Nonproliferation Treaty, 188
Nuclear Regulatory Commission, 67
Nuclear testing, 157
Nuclear waste disposal, 131, 135
Nuño, Guadalupe, 130
O'Brien, Donald, 57
O'Connor, John, 131–32

O'Dwyer, Jack, 53, 55
O'Leary, Hazel, 180, 237
Occupational diseases, 159
Occupational health and safety, 6, 21–22, 24, 158ff.
Occupational Health and Safety Act (1970), 33
Ocean dumping, 137–39
Office of Management and Budget (OMB), 67, 179, 191
Office of Technology Assessment, 135, 145
Oil and gas industry, 85, 183
Oil, Chemical, and Atomic Workers union, 159
Oil dependency, 45, 71
Ojibway, 172
Old-growth forests, 52, 189, 115, 212, 213, 214, 261
Oliver, Patsy, 130
OMB. *See* Office of Management and Budget
Option 9, 189–90
Oregon Natural Resource Council, 103, 193, 211, 213
Oren, Frank, 164
Orr, David, 216–17
Osborn, Henry Fairfield, 55, 90
Ospreys, 34
Overconsumption, 161, 165, 166,
Overpopulation. *See* Population control
Ozone depletion, 67, 70, 86, 88, 135, 184, 185, 259

Pacific Islanders, proximity to toxic-waste sites, 145
PACS. *See* Political action committees
PAHLS. *See* People Against Hazardous Landfill Sites
Passell, Peter, 90
PCBs (polychlorinated biphenyls), 7, 27, 33, 39
PCSD. *See* President's Council on Sustainable Development
Peace movement, 25, 239
Peng, Martin Khor Kok, 168
People Against Hazardous Landfill Sites (PAHLS), 130
People for Clean Air and Water. *See* El Pueblo para Aire y Agua Limpio
People for the West, 96
People of Color Network, 169, 208
People of Color Leadership Summit, 151–54, 156, 169, 223, 262
People of color, 153, 154–55. *See also* African Americans, Minorities, Native Americans, Hispanic Americans
Perkins, Jane, 186, 187, 199
Perrault, Michelle, 184, 237

Pesticides, 36, 131, 145, 256; data collection on illness related to, 145; deaths resulting from, 122; as carcinogen, 122; in Everglades, 183; federal pre-emption of state laws, 136; Hispanic farm worker exposure to, 155
Petrochemical industry, 71; support of anti-environmentalism, 84. *See also* Oil and gas industry
Petrochemicals-based economy, 111. *See also* Oil dependency
Pew Charitable Trust, 50–52, 103, 165
Philanthropy, 4, decline in environmental, 176; and fourth-wave environmentalism, 215; and mainstream nationals, 40 ff. *See also* Corporate philanthropy, Direct mail, Environmental philanthropy, Foundation philanthropy
Philosopy of environmentalism, 223
Pinchot, Gifford, 15–18, 37; at Boone and Crockett Club, 15; appointed by McKinley, 16; conflict with Muir, 16–18; friend of Theodore Roosevelt, 16; and resource management, 16–18; and scientific forestry, 15, 19
Pioneer Fund, 163
Planet Drum, 230
Political action committees (PACs), 85, 123, 192–93
Political Difficulties Facing Waste-to-Energy Conversion Plant Sitings (1984, Cerrell Report), 142
Pollot, Mark, 101
Polls, 3–4, 9–10, 32, 34, 43, 195–97, 260; and public support for environment, 72–73, 196; direct mail market test as, 46
Pollution, 4, 6, 7, 24; airborne, 7; "hot spots," 111; lead, 7; polychlorinated biphenyls (PCBs), 7, 27, 33, 39
Pollution credits (ERCs): in Clean Air Act amendments, 87; as gift of public resource, 123; electric utilities and, 113; invented by Dudek, 109; to reduce air pollution, 110; trade in, 110
Pollution Prevention Policy Statement (EPA), 117
Pollution tax, 114, 123
Polychlorinated biphenyls (PCBs): dumping in African-American community, 155; Memphis trash collectors' exposure to, 141. *See also* PCBs
Pomerance, Rafe, 119, 179
Pope, Carl, 186, 218
Population, 14, 160–66; Clinton policy on, 191; control of, 24, 71, 161, 229, 239; decline, 210; environmental agenda and, 161–62; future world, 160–61; North-South, 161
Population Bomb, 23, 161
Porter, Elliot, 20
Powell, John Wesley, 15
Power, Thomas Michael, 100, 123–24
PPPS. *See* Pollution Prevention Policy Statement (EPA)
Pragmatic reform, 59, 61. *See also* Culture of reform
Preservationism, 1, 8, 225; conflict with conservationism, 17; of early naturalists, 14; and Muir, 16–17; of Thoreau, 13
Preservationist-conservationist conflict, 16–18
President's Council on Sustainable Development (PCSD), 184, 237–38
Principles of Environmental Justice (*See* Appendix), 153
Private land, forestry practices on, 213, 218
Private property, 11, 78, 83, 91, 98–101. *See also* Takings
Progressive Greens, 242
"Promise, The." *See* Surface Mining Control and Reclamation Act (1977)
Property rights. *See* Takings
Proxmire, William, 59, 192
Public domain. *See* Public land
Public health legislation, 101
Public health movement, 20, 21–22, 24, 127
Public lands, 208, 212, 217, 251; cattle and mining on, 90; in early conservation movement, 19; logging abuse on, 103; under Pinchot, 16–18; and Sagebrush Rebellion, 90–92; and traditional environmental movement, 126; and Wise Use movement, 95
Public participation: grassroots insistence on, 135; N.C. biotechnology law and, 137; in NAFTA decisions, 185; third-wave discouragement of, 124
Public relations, corporate, 53, 55, 85;

Quayle, Dan, 108, 191

RACHEL. *See* Remote Access Chemical Hazards Electronic Library
Racism, 2, 143, 144ff., 156, 210, 253; at

EPA, 156; in environmental movement, 146, 152, 165, 166
Radioactive wastes, on Indian reservations, 145
Radon, 155
Rainforests, 45, 119, 194, 213
RARE. *See* Roadless Area and Review Evaluation
Rauber, Paul, 123–24
RCRA. *See* Resource Conservation and Recovery Act (1976), 33
Reagan administration, 6–8; anti-environmentalism of, 206; court appointments, 79, 214; halts EPA data collection on pesticide effects, 145; threat to wilderness, 68
Reagan, Ronald, 7–8, 11, 117; anti-environmentalism of, 65–68; commitment to material abundance, 66–67; court appointments of, 37, 99; and cost-benefit analysis, 56; election of, 89; executive order on private property, 98, 100; executive order on regulations, 101; exploitation of environmental backlash, 84; misperception of public opinion, 72–73; on regulatory flexibility, 106; as Sagebrush Rebel, 91; weakens EPA, 131
Reauthorizations, of environmental legislation, 154, 192, 194, 200
RECLAIM. *See* Los Angeles Regional Clear Air Incentives Program
Reform environmentalism, 29–61, 208–212; foundation influence, 49–53; and populist movements, 257; during Reagan years, 73. *See also* Culture of reform, Third-wave environmentalism
Regan, Richard, 57–58
Regulatory agencies, 243; and loss of democracy, 246; as subsidiaries of regulated industries, 64
Regulatory flexibility, 253; and Ronald Reagan, 106
Regulatory relief, 78
Rehnquist, William, 99–100
Reicher, Dan, 180
Reichert, Josh, 50–52, 103, 249–50
Reilly, William, 79–80, 107, 108, 123, 156, 179; appointed EPA administrator, 90; at Conservation Foundation, 89; and WMX incinerator, 113–14
Religion, and environmentalism, 232–33
Remote Access Chemical Hazards Electronic Library (RACHEL), 133
Renewable energy, 71
Rensi, Edward, 139–40
Resource Conservation and Recovery Act (1976, RCRA), 33, 58, 129, 143
Resource management: and Gifford Pinchot, 16–18; as element of environmentalism, 24
Resource Recovery and Conservation Act (RRCA), 65
Resources for the Future (RFF), 4, 55–56
Resources Recovery Act (1992), 154
RFF. *See* Resources for the Future
Rhodes, Deane, 91
Richardson, Bill, 58
Right to life, 244
Rimel, Rebecca, 52
Rio di Janeiro, 167, 182
Risk assessment, 181; in Clean Air Act amendments, 87–88; as environmental policy, 121; defined, 121–22. *See also* Cost-benefit analysis, Negligible risk
Roadless Area and Review Evaluation (RARE), 209
Roberts, Bill, 199
Rostenkowski, Dan, 183
Rockefeller Foundation, 239
Rockefeller, Laurance, 35, 90
Rodriguez, Richard, 220
Rooney, Philip, 57, 219
Roosevelt, Theodore, 16–17, 106
Roush, John, 235
Rousseau, Jean-Jacques, 12
Rubin, Robert, 178
Ruckelshaus, William, 33, 64, 123, 134, 156
Ruderman-Feuer, Gail, 105–106

Safe Drinking Water Act (1974), 33
Sagasti, Francisco, 206
Sagebrush Rebellion,67, 84, 90–92, 95
Sahara Club, 97
Sale, Kirkpatrick, 60, 140
San George, Robert, 56
Sand County Almanac, 18
Santa Barbara oil spill, 24, 128
Save the Manatee Club, 219
Save the Redwoods League, 2
Sawhill, John, 184, 186, 187, 188
Scaife, Richard Mellon, 84
Scalia, Antonin, 99
Scherr, Jacob, 119–20
Schlicheisen, Roger, 187
Schneiderman, Andres, 240
Science, environmental, 39–40, 108, 119

Scientific forestry, 15, 18
SCLDF. *See* Sierra Club Legal Defense Fund
Sea Shepherd Society, 211
SEACC. *See* Southeast Alaska Conservation Council
Seager, Joni, 231
Second Amendment Foundation, 93
Seed, John, 224, 227
Senate Environmental Affairs Committee, 86
Sewage systems, 21
"Sewergate," 65
Seymour, Whitney North, Jr., 35
Shabecoff, Philip, 222, 250, 262
"Shallow" ecology, 226–27
Shankar, Mani, 167
Sher, Vic, 48, 150, 169
Shifferle, Patti, 45
Sierra Accords, 215–16
Sierra Club, 2, 5, 30, 36, 74, 75, 86, 93, 94, 171, 182, 184, 186, 193, 209, 211, 261; in Ancient Forest Alliance, 213; ASCMEE referendum, 219; budget and membership decline of, 175, 247; and California timber industry, 115; in Clean Air Coalition, 87; compromises with timber industry, 214–16, 218; cuts programs, 247–48; democracy in, 216, 217–18; environmental justice program and, 169; Ethnic and Cultural Diversity Task Force, 164; as exemplification of environmentalism, 6; founded, 15; and Earth Day 1970, 25; and local activists, 214; dissident faction of, 216–18; Illinois Chapter of, 217; immigration and, 163–65; Kettleman City and, 149; local chapters of, 132; membership, budget growth in Reagan years, 70–71; and NAFTA, 187; and Oregon forests, 214; past racial policies of, 163; poor black communities and, 169; population issues and, 162, 164, 165; racial minorities in, 164–65; response to Sagebrush Rebellion, 92; response to Superfund site, 169; restructures, 250; and Sierra Accords, 215–16; and Wise Use movement, 101–102
Sierra Club Legal Defense Fund (SCLDF), 48, 169, 209, 213; breaks from Sierra Club, 36; founded, 35; Kettleman City and, 150; and old-growth forests, 52; and U.N. Human Rights Commission, 245;
Silent Spring, 1, 21, 23, 34, 84, 106
Sive, David, 35
SMCRA. *See* Surfacing Mining Control and Reclamation Act (1977)
Smith, Loren, 100
Snow, Donald, 92
Social ecology, 228–29
Social liberation, and wilderness preservation, 251
Social movement, 2–3, 8, 20, 28, 29, 205–206, 208
Soil loss, 236
Sosa, Marie, 130
South Dakota Sioux, 152
Southeast Alaska Conservation Council (SEACC), 136
Southern hemisphere: resource exploitation in, 120–21
"Stockholm Syndrome," 116
Southern Organizing Committee, 153–54
Southwest Network for Environmental and Economic Justice, 149,
Southwest Organizing Project, 146, 170
Special Project on Toxic Injustice (UCC), 157
Species extinction, 236
Speth, James Gustave "Gus," 35, 178
Spiritual ecology, 232–33
Spotted owls, 189, 190, 228
St. Clair, Jeffrey, 190–91
St. Louis Earth Day Committee, 27
Stevenson, Adlai, III, 64
Stone, I. F., 25
Storm King project, 35
Stringfellow Acid Co., 130
Strip mining, 79, 159, 183, 100
Strong, Ted, 237
Strontium 90, 39, 7, 33
Subsidies to timber industry, 212
Sulfur dioxide, 109, 111. *See also* Greenhouse gases
Superfund, 33, 63, 193; as defeat for corporations, 89; cleanup costs of, 89–90; economic impact of, 89; racial differences in implementation of, 143; Reagan attempts to subvert, 65; sites, 169; targeted by anti-environmentalists, 200
Superfund Act (1993), reauthorization of, 154
Superfund Coalition, 89–90
Sustainability, 26, 225; disagreements over,

235; extractive industries and, 235; in underevelopped countries, 236
Sustainable agriculture, 185
Sustainable development, 161, 162 , 235–38; as Native American ethic, 12; Powell's plan for, 15
Sustainable-yield forestry, 71, 115
Sutherland, Rick, 209
Syracuse University School of Forestry, 251
Szasz, Andrew, 129

Takings: 98–101; bills in Congress and state legislatures, 101; as grassroots backlash, 96, 101; definition of, 98; effect on toxics litigation, 101; Pollot's model bill, 101; Utah bill, 101
Tanton, John, 163
Task Force for Regulatory Relief, 68, 108
Task Force on Competition, 108
Taylor, Dorceta, 126, 146
Technologies: as value-free, 116; less-polluting, 109; environmental effects of, 20, 112–13
Third world countries: at Earth Summit, 166–67; effect of World Bank and IMF on, 118; environmentalists in, 120, 146; forest industries in, 119; population growth, and 163; oil development in, 119–20; resource exploitation in, 118
Third World Network, 168
Third-wave environmentalism, 8, 105–124, 181, 215; defined, 107–108; ebbing of, 206; effect on third world, 121; misleads public perceptions of pollution, 117; moral premise of, 123; PCSD rejects, 237
Thoreau, Henry David, 12, 13, 19, 60, 263
Thornburgh, Richard, 129
Thornton, James, 58
Three Ls, 78
3M Corporation, 112
Three-Mile Island, 129
Timber industry, 85, 212–14: and Ancient Forest Alliance, 213; and Forest Service, 212
Timber sales, 214, 261; in old-growth forests, 190–91
Toxic 500, 56
Toxic chemicals: bans or regulations of, 39–40
Toxic-control legislation: public demand for, 129
Toxic pollution, 6–7, 21–22, 226; of electric utilities, 110; reduction of, 109; in the workplace, 79, 158; *See also* Anti-toxics environmentalism
Toxic Release Inventory (EPA), 114
Toxic substances laws, effect of GATT on, 188
Toxic Substances Control Act (1976), 33
Toxic Ten list, 237
Toxic Waste and Race: A National Report on the Racial and Socioeconomic Characteristics of Communities with Hazardous-Waste Sites (1987), 142, 144
Toxic waste incinerators, 47, 114, 149–50, 179, 181
Toxic wastes, 129; cleanup of, 6, 89, 100; early campaigns against, 20–22; export of, 135, 147
Toxic-waste industry: anti-toxics movement and, 135; site selection of, 135, 142–143
Toxic-waste sites: and nonwhite communities, 141, 142
Toxics legislation, and takings, 101
Toxics Release Inventory, 22
Trade and environment, 263. *See also* General Agreement on Trade and Tariffs, North American Free Trade Agreement
Tradeoff, environmental/economic, 158, 159, 160
Train, Russell, 29, 123
Tree spiking, 210–11
Tripp, Jim, 138
TRIS, 36
Tundra Rebellion, 91
Turnage, Bill, 49, 69–70, 249

UCC. *See* United Church of Christ
U. S. Supreme Court, 99–100
U.S. Army Corps of Engineers, 138
U.S. Biological Survey, 30
U.S. Center for Disease Control, 145
U.S. Civil Rights Commission, 156
U.S. Congress, 60, 68, 180. *See also* Federal strategy, Environmental legislation, Environmental lobby
U.S. Council for Energy Awareness, 85
U.S. Court of Federal Claims, 98 ff.
U.S. Department of Agriculture, 16
U.S. Department of Defense, 130
U.S. Department of Energy, 67, 177
U.S. Department of Justice, 37
U.S. Department of Interior, 26, 100, 177
U.S Department of Labor, 79
U.S. Department of State, 180

U.S. Fish and Wildlife Service, 37, 179
U.S. Forest Service, 189, 209 213, 251, 261; captive of regulated industries, 18; established, 16; Leopold at, 19; as subsidiary of extractive industries, 64; subsidies to timber industry, 212; under Pinchot, 16, 18
U.S. General Accounting Office (GAO), studies hazardous-waste siting, 142
U.S. Greens, 229
U.S. immigration, 162–66
U.S. Office of Surface Mining, 183
Unfunded mandates, 81–82, 194
Union Carbide: in Superfund Coalition, 89
Union of Concerned Scientists, 74
United Church of Christ (UCC), 141–42, 155, 157
United Conference on Economic Development (Earth Summit), 166–67; North-South differences at, 166–67
United Fishermen, 138
United Nations Development Program, 178
United Nations Human Rights Commission, 244
Unleaded gasoline, 112
Uram, Robert, 183
Urban air standards, 86–87
Urban health problems, 21–22, 141, 145–46, 155
Urban League, 25, 171
Urban pollution, 20, 21–22
Utah takings bill, 101
Utilities, coal-burning: and ERCs, 110, 113
Utopia, 12

Valley of the Drums, 128
Vander Ryn, Sim, 235
Velsicol Chemical Company, 84–85
Vento, Bruce, 197
Verhoff, Catherine, 147
Viederman, Stephen, 236
Vietnam War, 25
Volatile organic chemicals (VOCs), 111
Voluntary compliance, 192
Voting, 196–97

W. Alton Jones Foundation, 50, 249
Wald, George, 63
Walinsky, Adam, 144–45
Walukas, Don, 107
War on poverty, 25
Ward, Justin, 185
Warren County, N.C., 141–42, 155
Washington Legal Foundation, 99
Waste-management industry, 63; costs of, 134; and NIMBYs, 130ff.; siting of facilities, 135, 142–43. *See also* Antipollution industry, Hazardous-waste handlers
Waste Management Inc. (WMX Technologies): and oil and gas cleanup, 58; EPA fines of, 55; executive on board of National Audubon, 55, 219; on NWF board, 153; oil-field waste regulation and, 219; toxic-waste incinerator of, 113–14;
Waste reduction, 135, 145, 188
WATCH. *See* Workers Against Toxic Chemicals
Water Pollution Control Act (1972), 33
Watson, Paul, 211
Watt, James, 11, 72, 91, 93, 177; appointed to Interior, 67; mainstream nationals target, 84, 92; resigns in disgrace, 65; transfers federal land to states, 92; violent rhetoric of, 97
Waxman, Henry, 199
Wayburn, Ed, 25
Weeden, Alan, 165
Weiss, Daniel J., 86
Western States Public Lands Coalition, 96
Wetlands, 96, 99, 195
Weyerhaeuser Co., 94
Wheeler, Douglas, 67
White House Council on Environmental Quality (CEQ). *See* Council on Environmental Quality
White House Office on Environmental Policy, 178
White phosphorous, worker exposure to, 21
Wild and Scenic Rivers Act (1968), 23
Wilderness Act (1964), 23, 31, 32
Wilderness, concept of, 11
Wilderness preservation, 2, 6, 8, 10, 16–18, 24, 68, 126, 153, 170, 174, 217
Wilderness Society, 30, 69, 76 179, 193; 209, 211, 249, 251; in Ancient Forest Alliance, 213; budget and membership decline of, 175; and Earth Day 1970, 25; growth in membership, budgets in Reagan years, 70–71; response to Sagebrush Rebellion, 92
Wildlife protection, 2, 6, 13, 106, 153, 170